SpringerWienNewYork

CISM COURSES AND LECTURES

The series presents lecture notes, monographs, edited works and proceedings in the field of Mechanics, Engineering, Computer Science and Applied Mathematics.
Purpose of the series is to make known in the international scientific and technical community results obtained in some of the activities organized by CISM, the International Centre for Mechanical Sciences.

INTERNATIONAL CENTRE FOR MECHANICAL SCIENCES

COURSES AND LECTURES - No. 538

MULTIPHASE MICROFLUIDICS: THE DIFFUSE INTERFACE MODEL

EDITED BY

ROBERTO MAURI
UNIVERSITY OF PISA, ITALY

SpringerWienNewYork

This volume contains 35 illustrations

SPIN 86085134

All contributions have been typeset by the authors.

ISBN 978-3-7091-1226-7 SpringerWienNewYork

PREFACE

The aim of this text is to review the theory and describe some relevant applications of the interface diffuse model for one-component, two-phase uids and for liquid binary mixtures, to model multiphase ows in con ned geometries.

In classical models, multiphase ows are described assuming that the different phases are separated by a sharp interface, with appropriate boundary conditions. Imposing that the condition of local equilibrium is satis ed, physical properties such as density and composition are allowed to change discontinuously across the interface, so that one of the main di culties of this approach consists in solving the free boundary problem and determine the position of the interface. In fact, interface tracking breaks down whenever the real interface thickness is comparable with the lengthscale of the phenomenon that is being studied, as it happens, naturally, in micro-devices. This problem arises elsewhere as well, such as in modeling drop coalescence and break up and in describing the wetting and de-wetting of solid surfaces. In these cases, it is more reasonable to use a different approach, as was proposed at the end of the 19th century by Rayleigh and Van der Waals, where interfaces have a non-zero thickness, i.e. they are diffuse, so that all quantities can vary continuously.

In the rst chapter of this text, Lamorgese, Molin and Mauri review extensively the diffuse interface approach. The equations of motion are derived, showing how additional stresses, the so called Korteweg stresses, arise naturally as reversible body forces that tend to minimize the free energy of the system, acting exclusively in the interface region. Several case studies will be presented to demonstrate the advantages of the diffuse interface method in modeling multiphase ows in microdevices, as compared to the classical two-phase ow approach. Among the examples that will be presented, here we mention the following problems: a) mixing, spinodal decomposition and nucleation of macroscopically quiescent regular binary liquid mixtures; b) vapor-liquid phase separation of a single component uid; c) heat transfer enhancement due to phase change; g) spontaneous emergence of complex structures during growth far from equilibrium.

This model is further extended in the second chapter by Park, Mauri and Anderson, to analyze the phase separation of regular ternary *liquid mixtures. Unlike critical binary mixtures, which during spinodal decomposition evolve into bicontinuous structures, ternary mixtures seem to loose such symmetry and therefore exhibit different scalings. A considerable emphasis is devoted to the advanced numerical modeling schemes that have been developed so far, stressing the computational di culties encountered in implementing the diffuse interface method.*

The third chapter consists of Thiele's account of the continuum approaches to thin lms of liquid mixtures and, in particular, of recent studies of dewetting lms for simple and complex liquids. After describing the basic dewetting models for one-layer single-phase liquid lms, the case of binary mixtures undergoing both dewetting and decomposition processes is discussed, assuming that the lms rst decompose into strati ed lms and then evolve into lateral structures. Two approaches are presented: a two-layer sharp-interface theory in the form of coupled evolution equations for the layer thickness pro les and a diffuse-interface short-wave one-domain model, with boundary conditions at the sharp liquid-solid and liquid-gas interfaces. After describing the linear stability of strati ed lms, advantages and disadvantages of the diffuse interface model are analyzed as compared to the sharp interface approach.

Finally, in the forth chapter, Plapp generalizes the diffuse interface approach to cases in which the phase eld cannot be identi ed (as in the diffuse interface approach) with a physical quantity (coarse-grained on a mesoscopic scale), and instead can only be interpreted as a smoothed indicator function. In particular, a phenomenological phase- eld model for solidi cation is introduced, showing that, with a proper choice of some interpolation functions, surface and bulk properties can be adjusted independently. The link between this phase- eld model and the classic free-boundary formulation of solidi cation is established by the use of matched asymptotic analysis. As examples of applications of this approach, the solidi cation of alloys and the advected eld model for two-phase ows are brie y discussed.

The text is addressed to doctoral students, young researchers as well as practicing R&D engineers, who are interested in micro uidics and, in general, with multiphase ows.

Roberto Mauri

CONTENTS

Diffuse Interface (D.I.) Model for Multiphase Flows
by A.G. Lamorgese, D. Molin and R. Mauri 1

Phase Separation of Viscous Ternary Liquid Mixtures
by J.M. Park, R. Mauri and P.D. Anderson 73

Dewetting and Decomposing Films of Simple and Complex Liquids
by U. Thiele 93

Phase-Field Models
by M. Plapp 129

Di use Interface (D.I.) Model for Multiphase Flows

Andrea G. Lamorgese, * Dafne Molin † and Roberto Mauri

† Department of Chemical Engineering, The City College of CUNY, 10031 New York, U.S.A.

† Department of Energy Technology, University of Udine, 33100 Udine, Italy.

‡ Department of Chemical Engineering, Industrial Chemistry and Material Science, University of Pisa, 56126 Pisa, Italy.

Abstract

We review the diffuse interface model for fluid flows, where all quantities, such as density and composition, are assumed to vary continuously in space. This approach is the natural extension of van der Waals' theory of critical phenomena both for one-component, two-phase fluids and for liquid binary mixtures. The equations of motion are derived, showing that the problem is well posed, as the rate of change of the total energy equals the energy dissipation. In particular, we see that a non-equilibrium, reversible body force appears in the Navier-Stokes equation, that is proportional to the gradient of the generalized chemical potential. This, so called Korteweg, force is responsible for the convective motion observed in otherwise quiescent systems during phase change. Finally, the results of several numerical simulations are described, modeling, in particular, a) mixing, b) spinodal decomposition; c) nucleation; d) heat transfer; e) liquid-vapor phase separation.

1 Introduction

The theory of multiphase systems was developed at the beginning of the 19th century by Young, Laplace and Gauss, assuming that different phases are separated by an interface, that is a surface of zero thickness. In this approach, physical properties such as density and concentration, may change discontinuously across the interface and the magnitude of these jumps can be determined by imposing equilibrium conditions at the interface. For example, imposing that the sum of all forces applied to an infinitesimal curved interface must vanish leads to the Young-Laplace equation, stating that the difference in pressure between the two sides of the interface (where each phase is assumed to be at equilibrium) equals the product of surface tension and curvature. Later, this approach was generalized by defining surface thermodynamical properties, such as surface energy and entropy, and surface transport quantities, such as surface viscosity and heat conductivity, thus formulating the thermodynamics and transport phenomena of multiphase systems. At the end of the 19th century, though, another approach was proposed by Rayleigh (1892) and van der Waals (1893), who assumed that interfaces have a non-zero thickness, i.e. they are "diffuse." Actually, the basic idea was not new, as it dated back to Maxwell (1876) and Gibbs (1876), Poisson (1831) and von Leibnitz (1765) or even Lucretius (50 B.C.E.), who wrote that "a body is never wholly full nor void." Concretely, in a seminal article published in 1893, van der Waals (1893) used his equation of state to predict the thickness of the interface, showing that it becomes infinite as the critical point is approached. Later, Korteweg (1901) continued this work and proposed an expression for the capillary stresses, which are generally referred to as Korteweg stresses, showing that they reduce to surface tension when the region where density changes from one to the other equilibrium value collapses into a sharp interface (see Rowlinson and Widom, 1982, for a review of the molecular basis of capillarity).

In the first half of the 20th century, van der Waals' theory of critical phenomena was generalized by Ginzburg and Landau (Landau and Lifshitz, 1980), leading to a general theory of second-order phase transition and thereby describing phenomena such as ferromagnetism, superfluidity and superconductivity. Then, at mid 1900, Cahn and Hilliard (1959) applied van der Waals' diffuse interface (D.I.) approach to binary mixtures and then used it to describe nucleation and spinodal decomposition (Cahn, 1961). This approach was later extended to model phase separation of polymer blends and alloys (de Gennes, 1980, and references therein). Finally, in the mid 1970s, the D.I. approach was coupled to hydrodynamics, developing a set of conservation equations, thanks to the work by, among others, Kawasaki (1970), Siggia (1979), and Hohenberg and Halperin (1977). These latter authors referred to this approach as "model H" and only later the name "diffuse interface method" was introduced. Finally, recent developments in computing technology have stimulated a resurgence of the D.I. approach, above all in the study of systems with complex morphologies and topological changes. A detailed discussion about D.I. theory coupled with hydrodynamics can be found in Antanovskii (1996), Lowengrub and Truskinovsky (1998), Anderson et al. (1998) and, more recently, in Onuki (2007) and Thiele et al. (2007). In order to better understand the basic idea underlying the D.I. theory, let us remind briefly the classical approach to multiphase flow that is used in fluid mechanics. There, the equations of conservation of mass, momentum, energy and chemical species are written separately for each phase, assuming that temperature, pressure, density and composition of each phase are equal to their equilibrium values. Accordingly, these equations are supplemented by boundary conditions at the interface, namely (Davis and Scriven, 1982),

$$[\![\tau]\!]_+^- \cdot \mathbf{n} = \kappa\sigma\mathbf{n} + (\mathbf{I} - \mathbf{nn}) \cdot \nabla\sigma, \qquad [\![\mathbf{v}]\!]_+^- = 0, \qquad [\![T]\!]_+^- = 0, \tag{1}$$

with $\mathbf{n}$ denoting the normal at the interface, stating that the jump of the stress tensor at the interface is related to the the curvature κ, the surface tension σ and its gradient, while velocity $\mathbf{v}$ and temperature T are continuous (unless we introduce concepts such as surface viscosity and surface heat conductivity, so that they become discontinuous as well). Naturally, this results in a free boundary problem, which means that one of the main problems of this approach is to determine the position of the interface. To that extent, many interface tracking methods have been developed, which have proved very successful in a wide range of situations. However, interface tracking breaks down whenever the interface thickness is comparable to the length scale of the phenomenon that is being studied, such as a) in near-critical fluids or partially miscible mixtures, as the interface thickness diverges at the critical point, and the morphology of the systems presents self-intersecting free boundaries; b) near the contact line along a solid surface, in the breakup/coalescence of liquid droplets and, in general, in microfluidics, as the related physical processes act on length scales that are comparable to the interface thickness. In front of these difficulties, the D.I. method offers an alternative approach. Quantities that in the free boundary approach are localized in the interfacial surface, here are assumed to be distributed within the interfacial volume. For example, surface tension is the result of distributed stresses within the interfacial region, which are often called capillary, or Korteweg, stresses. In general, the interphase boundaries are considered as mesoscopic structures, so that any material property varies smoothly at macroscopic distances along the interface, while the gradients in the normal direction are steep. Accordingly, the main characteristic of the D.I. method is the use of an "order parameter" which undergoes a rapid but continuous variation across the interphase boundary, while it varies smoothly in each bulk phase, where it can even assume constant equilibrium values. For a single-component system, the order parameter is the fluid density ρ, for a liquid binary mixture it is the molar (or mass) fraction ϕ, while in other cases it can be any other parameter, not necessarily with any physical meaning, that allows to reformulate free boundary problems. In all these cases, the D.I. model must include a characteristic interface thickness, over which the order parameter changes. In fact, in the asymptotic limit of vanishing interfacial width, the diffuse interface model reduces to the classical free boundary problem. In Section 2 and 3 we formulate the diffuse interface model for single-component fluids and liquid binary mixtures at equilibrium, respectively. Then, in Section 4, the equations of motion are developed for non-dissipative systems, while in Section 5 these results are generalized to dissipative systems. Finally, in Section 7, results of numerical simulations are presented for the case of regular liquid binary mixtures and single-component van der Waals fluids.

2 Equilibrium conditions for single-component fluids

2.1 The free energy and van der Waals' equation

All thermodynamical properties can be determined from the Helmholtz free energy. This, in turn, depends on the intermolecular forces which, in a dense fluid, are a combination of weak and strong forces. Fortunately, strong interactions nearly balance each other, so that the net forces acting on each molecule are weak and long-range. In addition, mean field approximation is assumed to be applicable, meaning that molecular interactions are smeared out and can be replaced by the action of a continuous effective medium (Pismen, 2001). Based on these assumptions, the case of dense fluids can be treated as that of nearly ideal gases, so that, allowing for variable density, the molar Helmholtz free energy at constant temperature T can be written as (Landau and Lifshitz, 1980, Ch. 74):

$$f[\rho(\mathbf{x})] = f_{id} + \frac{1}{2}RTN_A \int \left(1 - e^{-U(r)/kT}\right) \rho(\mathbf{x}+\mathbf{r})\, d^3\mathbf{r}, \tag{2}$$

where k is Boltzmann's constant, $R = N_A k$ is the gas constant, with N_A the Avogadro number, U is the pair interaction potential, which depends on the distance $r = |\mathbf{r}|$, ρ is the molar density, while the factor $1/2$ compensates counting twice the interacting molecules. The first term on the RHS,

$$f_{id} = RT \ln \rho, \tag{3}$$

is the molar free energy of an ideal gas (where molecules do not interact). Now, we assume that the interaction potential consists of a long-range term, decaying as r^{-6} (like in the Lennard-Jones potential), while the short-range term is replaced by a hard-core repulsion, i.e. (Israelachvili, 1992),

$$U(r) = \begin{cases} -U_0(r/l)^6 & (r > d) \\ \infty & (r < d) \end{cases} \tag{4}$$

where d is the nominal hard-core molecular diameter, l is a typical intermolecular interaction distance, and the non-dimensional constant U_0 represents the strength of the intermolecular potential. When the density is constant, Eq. (2) gives the thermodynamic free energy, f_{Th},

$$f_{Th}(T,\rho) = f_{id}(T,\rho) + f_{ex}(T,\rho), \tag{5}$$

where

$$f_{ex}(T,\rho) = RT\rho B(T), \tag{6}$$

is the excess (i.e. the non ideal part) of the free energy, with

$$B(T) = \frac{1}{2}N_A \int_0^\infty (1 - e^{-U(r)/kT}) 4\pi r^2 \, dr \tag{7}$$

denoting the first virial coefficient. This integral can be solved as

$$B(T) = 2\pi N_A \int_0^d r^2 dr + 2\pi N_A \int_d^\infty (1 - e^{\frac{U_0}{kT}(r/l)^6}) r^2 \, dr = c_2 - \frac{c_1}{RT}, \tag{8}$$

where

$$c_1 = \frac{2}{3}\pi U_0 N_A^2 l^6/d^3 \quad \text{and} \quad c_2 = \frac{2}{3}\pi d^3 N_A \tag{9}$$

are the pressure adding term and the excluded molar volume, respectively. Finally we obtain:

$$f_{Th}(\rho,T) = f_{id} + RTc_2\rho - c_1\rho \approx RT\ln\left(\frac{\rho}{1-c_2\rho}\right) - c_1\rho, \tag{10}$$

that is

$$f_{Th}(\rho,T) = -RT\ln(v - c_2) - \frac{c_1}{v}, \tag{11}$$

where $v = \rho^{-1}$ is the molar volume. At this point, applying the thermodynamic equality (Landau and Lifshitz, 1980, Ch. 76) $P = -(\partial f/\partial v)_{N,T}$, we obtain the van der Waals' equation of state

$$\left(P + \frac{c_1}{v^2}\right) = \frac{RT}{v - c_2}. \tag{12}$$

This equation of state could be considerably improved if the term $RT/(v-c_2)$, which is exact in one dimension, is replaced by a more accurate representation of the pressure for a hard-sphere fluid in three dimensions (Widom, 1996).

2.2 The critical point

In the $P-T$ diagram, the vapor-liquid equilibrium curve stops at the critical point, characterized by a critical temperature T_C and a critical pressure P_C. At higher temperatures, $T > T_C$, and pressures, $P > P_C$, the differences between liquid and vapor phases vanish altogether and we cannot even speak of two different phases. In particular, as the critical point is approached, the difference between the specific volume of the vapor phase and that of the liquid phase decreases, until it vanishes at the critical point. Accordingly, near the critical point, since the specific volumes of the two phases, v and $v+\delta v$, are near to each other, we obtain:

$$P(T,v) = P(T,v+\delta v) = P(T,v) + \left(\frac{\partial P}{\partial v}\right)_T \delta v + \frac{1}{2}\left(\frac{\partial^2 P}{\partial v^2}\right)_T (\delta v)^2 + \ldots,$$

where we have considered that the two phases at equilibrium have the same pressure, in addition to having the same temperature. At this point, dividing by v and letting $\delta v \to 0$, we see that at the critical point we have:

$$\left(\frac{\partial P}{\partial v}\right)_T = 0, \quad \text{that is,} \quad \kappa_T \to \infty \text{ as } T \to T_C, \tag{13}$$

where κ_T is the isothermal compressibility. Note that this condition is the limit case of the inequality $(\partial P/\partial v)_T \leq 0$, which manifests the internal stability of any single-phase system. In addition, since near an equilibrium point, $\delta f + P\delta v > 0$, expanding δf in a power series of δv, with constant T, we obtain:

$$\delta f_{Th} = \left(\frac{\partial f_{Th}}{\partial v}\right)_T (\delta v) + \frac{1}{2!}\left(\frac{\partial^2 f_{Th}}{\partial v^2}\right)_T (\delta v)^2 + \frac{1}{3!}\left(\frac{\partial^3 f_{Th}}{\partial v^3}\right)_T (\delta v)^3 + \ldots$$

Finally, considering that $(\partial f_{Th}/\partial v)_T = -P$ and that at the critical point $(\partial^2 f_{Th}/\partial v^2)_T = 0$, we obtain:

$$\frac{1}{3!}\left(\frac{\partial^2 P}{\partial v^2}\right)_{T_C} (\delta v)^3 + \frac{1}{4!}\left(\frac{\partial^3 P}{\partial v^3}\right)_{T_C} (\delta v)^4 + \ldots < 0.$$

Since this equality must be valid for any value (albeit small) of δv (both positive and negative), we obtain:

$$\left(\frac{\partial^2 P}{\partial v^2}\right)_{T_C} = 0, \quad \left(\frac{\partial^3 P}{\partial v^3}\right)_{T_C} < 0. \tag{14}$$

Therefore, the critical point corresponds to a horizontal inflection point in the $P - v$ diagram, which means that, since $P = -(\partial f_{Th}/\partial v)_T$,

$$\left(\frac{\partial^2 f_{Th}}{\partial v^2}\right)_{T_C} = 0, \quad \left(\frac{\partial^3 f_{Th}}{\partial v^3}\right)_{T_C} = 0. \tag{15}$$

Imposing that at the critical point the $P - v$ curve has a horizontal inflection point, we can determine the constant c_1 e c_2 in the van der Waals equation (the same is true for any two-parameter cubic equation of state) in terms of the critical constant T_C and P_C, finding (Landau and Lifshitz, 1980, Ch. 84):

$$c_1 = \frac{9}{8}RT_C v_C = \frac{27}{64}\frac{(RT_C)^2}{P_C} \quad \text{and} \quad c_2 = \frac{1}{3}v_C = \frac{1}{8}\frac{RT_C}{P_C}. \tag{16}$$

Viceversa, the critical pressure, temperature and volume can be determined as functions of c_1 and c_2 as follows:

$$P_C = \frac{1}{27}\frac{c_1}{c_2^2}, \quad T_C = \frac{8}{27}\frac{c_1}{Rc_2}, \quad v_C = 3c_2. \tag{17}$$

Using these expressions, the van der Waals equation can be written in terms of the reduced coordinates as:

$$\left(P_r + \frac{3}{v_r^2}\right)(3v_r - 1) = 8T_r, \quad P_r = \frac{P}{P_C}, \quad v_r = \frac{v}{v_C}, \quad T_r = \frac{T}{T_C}. \tag{18}$$

This equation represents a family of isotherms in the $P_r - v_r$ plane describing the state of any substance, which is the basis of the law of corresponding states. As expected, when $T_r > 1$ the isotherms are monotonically decreasing, in agreement with the stability condition $(\partial P/\partial v)_T < 0$, while when $T_r < 1$ each isotherm has a maximum and a minimum point and between them we have an instability interval, with $(\partial P/\partial v)_T > 0$, corresponding to the two-phase region (see Figure 1).

Note that, considering that $P_C v_C = (3/8)RT_C$ and substituting the expressions for c_1 and c_2 in terms of the intermolecular potential, we obtain the following relation:

$$\left(\frac{l}{d}\right)^2 = \frac{3}{2}\left(\frac{kT_C}{U_0}\right)^{1/3}. \tag{19}$$

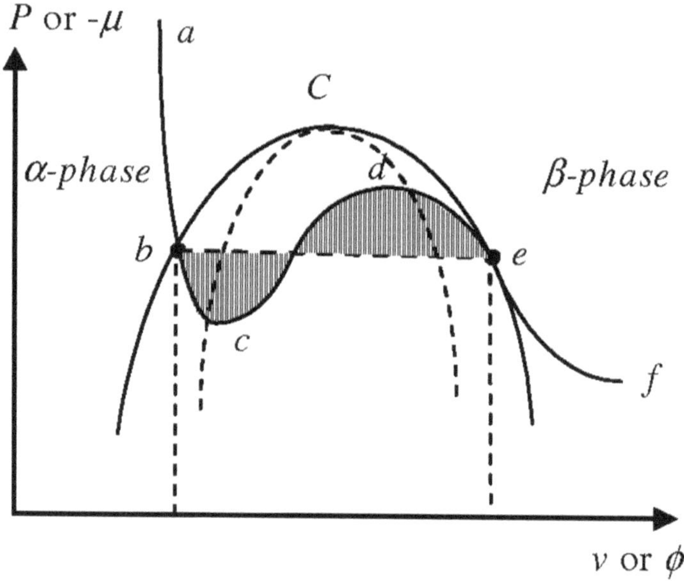

Figure 1. Phase diagram (P vs. v) of a single component fluid and ($-\mu$ vs. ϕ) of a binary mixture.

2.3 Coexistence and spinodal curves

Let us consider a one-component system at equilibrium, whose pressure and temperature are below their critical values, so that it is separated into two coexisting phases, say α and β. According to the Gibbs phase rule, these two phases have the same pressure and temperature and therefore, defining the Gibbs molar free energy $g_{Th} = f_{Th} + Pv$, with $dg_{Th} = -sdT + vdP$, the corresponding equilibrium, or saturation, pressure P_{sat} at a given temperature can be easily determined from the equilibrium condition, stating that at equilibrium the Gibbs molar free energies of the two phases must be equal to each other. So we obtain:

$$g_{Th}^{\beta} - g_{Th}^{\alpha} = \int_b^e dg_{Th} = 0 \quad \Longrightarrow \quad \int_b^e v\, dP = [vP]_b^e - \int_b^e P\, dv = 0, \quad (20)$$

where $P = P(v)$ represents an isotherm transformation. From a geometrical point of view, this relation manifests the equality between the shaded area of Figure 1 (Maxwell's rule), where the points b and e correspond to the equilibrium, or saturation, points of the liquid and vapor phases at that temperature at equilibrium, respectively, with specific volumes v_e^{α} and v_e^{β}. Conversely, the specific volumes of the two phases at equilibrium could also be determined from the molar free energy f_{Th}, rewriting Eq. (18) in terms of reduced coordinates as

$$\frac{f_{Th}}{RT_C} = -T_r \ln(v_C) - T_r \ln\left(v_r - \frac{1}{3}\right) - \frac{9}{8v_r}. \quad (21)$$

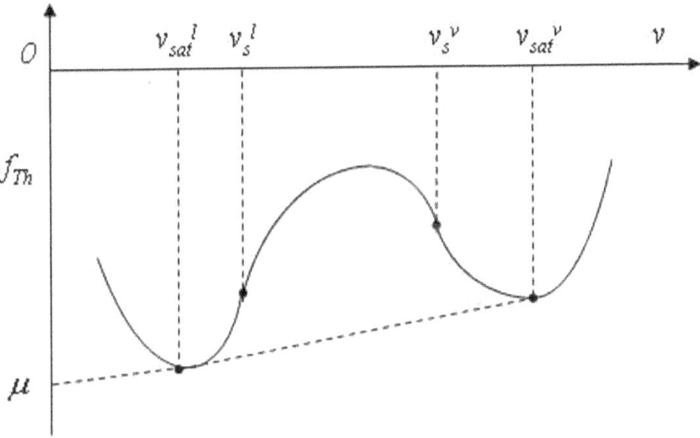

Figure 2. Typical double-well curve of the free energy of a single component fluid.

When $T_r < 1$ a typical curve of the free energy is represented in Figure 2. Now, keeping T_r fixed and considering that the two phases at equilibrium have the same pressure, using the relation $P = -(\partial f_{Th}/\partial v)_T$, we obtain:

$$P^\alpha = P^\beta \quad \Longrightarrow \quad \left(\frac{\partial f_{th}}{\partial v}\right)_T^\alpha = \left(\frac{\partial f_{Th}}{\partial v}\right)_T^\beta, \tag{22}$$

which, in Figure 2, represents the fact that the two equilibrium points have the same tangent. From this relation we can determine the specific volumes of the two phases at equilibrium, v_e^α and v_e^β. This relation can also be obtained considering that the specific volumes of the two phases at equilibrium minimize the total free energy, i.e.,

$$F_{Th} = \int \hat{f}_{Th}(\rho)\, d^3\mathbf{x} = min., \tag{23}$$

where $\hat{f} = \rho f$ is the free energy per unit volume,

$$\hat{f}_{Th} = \rho f_{Th} = \hat{f}_{id} + \hat{f}_{ex} = RT[\rho \ln \rho + \rho^2 B(T)]. \tag{24}$$

This minimization is carried out in Section 2.6. In Figure 1, besides the equilibrium curve, we have represented the, so called, spinodal curve, defined as the locus of all points (like c and d) satisfying $(\partial P/\partial v)_T = 0$. When the equilibrium and spinodal points are plotted in a $T - v$ diagram, we obtain the curves of Figure 3. All points lying outside the region encompassing the equilibrium curve are stable and represent homogeneous, single-phase systems; all points lying inside the region within the bell-shaped spinodal curve are unstable and represent systems that will separate into two phases (one liquid and another vapor, in this case); the region sandwiched between the equilibrium and the spinodal curves represents metastable systems, that is overheated liquid and undercooled vapor. The spinodal points can be also determined using the relation $(\partial P/\partial v)_T = 0$, obtaining:

$$\left(\frac{\partial^2 f_{Th}}{\partial v^2}\right)_T = 0, \tag{25}$$

determining the spinodal specific volumes $\tilde{v}_s^\alpha$ and $\tilde{v}_s^\beta$.

2.4 The critical exponents

Let us turn now to study the equation of state of a single component system close to its critical point. Then, instead of T, P and v, it is convenient to use the following variables:

$$\tilde{t} \equiv T_r - 1 = \frac{T - T_C}{T_C}, \quad \tilde{p} \equiv P_r - 1 = \frac{P - P_C}{P_C}, \quad \tilde{v} \equiv v_r - 1 = \frac{v - v_C}{v_C}, \tag{26}$$

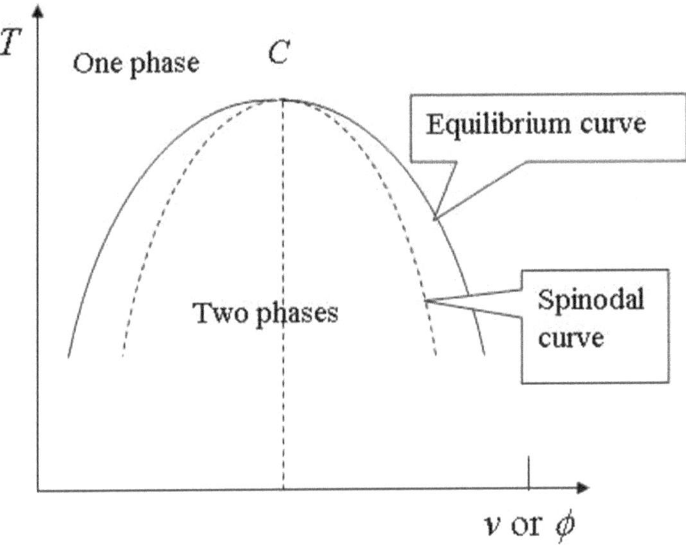

Figure 3. Phase diagram $(T - v)$ of a single component fluid and $(T - \phi)$ of a binary mixture.

where $T_r = T/T_C$, $P_r = P/P_C$ and $v_r = v/v_C$ are the reduced variables. At the end of a tedious, but elementary power expansion in terms of these variables, we see that the van der Waals equation reads, neglecting higher order terms,

$$\tilde{p} = 4\tilde{t} - 6\tilde{t}\tilde{v} - \frac{3}{2}\tilde{v}^3. \tag{27}$$

Note that we cannot have any term proportional to $\tilde{v}$ or $\tilde{v}^2$, in agreement with the conditions $(\partial P/\partial \tilde{v})_{T_C} = (\partial^2 P/\partial \tilde{v}^2)_{T_C} = 0$, while the coefficient of the $\tilde{v}^3$-term must be negative, as $(\partial^3 P/\partial \tilde{v}^3)_{T_C} < 0$. When $\tilde{t} > 0$, all states of the system are stable, that is there is no phase separation and the system remains homogeneous. This means that, when $\tilde{t} > 0$, it must be $(\partial P/\partial \tilde{v})_{T_C} < 0$, and therefore the coefficient of the $\tilde{t}\tilde{v}$-term must be negative. Finally, note that the $\tilde{t}\tilde{v}^2$ and $\tilde{t}^2\tilde{v}$ terms have been neglected because they are much smaller than $\tilde{t}\tilde{v}$, while the $\tilde{t}\tilde{v}$-term must be kept, despite being much smaller than $\tilde{t}$, for reasons that will be made clear below. In general, in the vicinity of the critical point the isotherms of a homogeneous system can be written as

$$\tilde{p} = b_c\tilde{t} - 2a_c\tilde{t}\tilde{v} - 4B_c\tilde{v}^3. \tag{28}$$

This expression for the free energy is the basis of Landau's mean field theory (Landau and Lifshitz, 1980, Ch. 146 and 148); it corresponds to van der Waals' Eq. (27) with $a_c = 3, b_c = 4$ and $B_c = 3/8$. At the critical point, where $\tilde{t} = 0$, from (28) we obtain: $\tilde{p} \propto \tilde{v}^\delta$, where $\delta = 3$ is a critical exponent, that we find unaltered in all critical phenomena. Following an isotherms $\tilde{t} = constant$, in the unstable region between the spinodal points, where $(\partial P/\partial v)_T > 0$, the system separates into two coexisting phases. At equilibrium, the two specific volumes satisfy Eq. (22), and considering that

$$\left(\frac{\partial \tilde{p}}{\partial \tilde{v}}\right)_{\tilde{t}} = -2a_c\tilde{t} - 12B_c\tilde{v}^2, \tag{29}$$

we can determine the specific volumes of the two phases at equilibrium:

$$\tilde{v}_e^\alpha = -\tilde{v}_e^\beta = \sqrt{\frac{-a_c\tilde{t}}{2B_c}} \quad \Longrightarrow \quad \Delta\tilde{v}_e^{\alpha\beta} = \tilde{v}_e^\alpha - \tilde{v}_e^\beta \propto (-\tilde{t})^\beta, \tag{30}$$

where $\beta = 1/2$ is another critical exponent. Now we see why we could not neglect the $\tilde{v}\tilde{t}$ term: if we did, the two specific volumes would result equal to each other. Note that the difference between the specific volumes of points that lie on the spinodal curve can be determined as well, applying the condition $(\partial\tilde{p}/\partial\tilde{v})_{\tilde{T}} = 0$, obtaining:

$$\tilde{v}_s^\alpha = -\tilde{v}_s^\beta = \sqrt{\frac{-a_c\tilde{t}}{6B_c}}. \tag{31}$$

The critical properties can also be determined from the free energy f_{Th}. In fact, integrating $(df_{Th})_T = -Pdv$ and substituting (28), we see that in the vicinity of the critical point the free energy has the following expression:

$$f_{Th}(\tilde{v}, \tilde{t}) = P_C v_C[h(\tilde{t}) + (1 + b_c\tilde{t})\tilde{v} + a_c\tilde{t}\tilde{v}^2 + B_c\tilde{v}^4], \tag{32}$$

where $h(\tilde{t})$ is an undetermined function of the temperature. For a van der Waals system, we obtain the same result expanding (21); in that case, $h(\tilde{t}) = (1 + \tilde{t})\ln[3/(2v_C)] - 9/8$. Finally, applying (22) and (25), we obtain again (30) and (31). Another critical exponent, γ, is defined as $\kappa_T^{-1} \propto \tilde{t}^\gamma$, where κ_T is the isothermal compressibility coefficient. From its definition, we obtain:

$$\kappa_T^{-1} = -v\left(\frac{\partial P}{\partial v}\right)_T = -P_C(1+\tilde{v})\left(\frac{\partial \tilde{p}}{\partial \tilde{v}}\right)_{\tilde{T}} \cong 2a_c P_C \tilde{t}, \tag{33}$$

showing that $\gamma = 1$. On the equilibrium curve, with $\tilde{t} < 0$ and $\tilde{v} = 0$, we have $\tilde{p} = b_c\tilde{t}$ [see Eq. (28)]. Therefore, applying the Clausius-Clapeyron equation,

$$\left(\frac{dP}{dT}\right)^{sat} = \frac{\Delta h^{\alpha\beta}}{T(v^\beta - v^\alpha)} \quad \Longrightarrow \quad \frac{d\tilde{P}}{d\tilde{t}} = b_c \cong \frac{\Delta h^{\alpha\beta}}{P_C v_C(\tilde{v}_e^\beta - \tilde{v}_e^\alpha)}, \tag{34}$$

where $\Delta h^{\alpha\beta}$ is the latent heat of vaporization, we obtain an expression for $\Delta h^{\alpha\beta}$ near the critical point,

$$\Delta h^{\alpha\beta} \cong b_c R T_C(\tilde{v}^\beta - \tilde{v}^\alpha) = \sqrt{\frac{a_c b_c^2}{B_c}} R T_C \sqrt{-\tilde{t}}, \tag{35}$$

where we have substituted Eq. (30), considering that $P_C V_C \cong RT_C$. From this equation we see that on approaching the critical point the latent heat of evaporation vanishes like $\sqrt{-\tilde{t}}$.

Finally, the last thermodynamic quantity to be determined is the specific heat. From the definition, $c_v = (\partial u_{Th}/\partial T)_V$, where u_{Th} is the molar internal energy, and considering that $u_{Th} = f_{Th} + Ts = f_{Th} - T(\partial f_{Th}/\partial T)_v$, we see that $c_v = -T(\partial^2 f_{Th}/\partial T^2)_V$. Therefore, from Eq. (32), it appears that c_v^{sat} remains finite at the critical point, that is $c_v \propto (-\tilde{t})^{-\alpha}$, where $\alpha = 0$ is another critical exponent. Consequently, using the well known relation

$$c_p - c_v \propto \left[\left(\frac{\partial \tilde{p}}{\partial \tilde{t}}\right)_{\tilde{v}}\right]^2 \bigg/ \left(\frac{\partial \tilde{p}}{\partial \tilde{v}}\right)_{\tilde{t}}, \tag{36}$$

we see that, since $(\partial\tilde{p}/\partial\tilde{t})_{\tilde{t}=0,\tilde{v}=0} = b_c$ and $(\partial\tilde{p}/\partial\tilde{v})_{\tilde{t}=0,\tilde{v}=0} = 0$, the specific heat c_p diverges. In fact we find:

$$c_p \propto \frac{1}{(\partial\tilde{p}/\partial\tilde{v})_{\tilde{t}}} = \frac{1}{-2a_c\tilde{t} - 12B_c\tilde{v}^2} \propto \frac{1}{(-\tilde{t})}, \tag{37}$$

where we have considered that on the equilibrium curve, $\tilde{v} \propto \sqrt{-\tilde{t}}$.

There are two more critical exponents. The first describes the vanishing of the surface tension σ in the vicinity of the critical point, $\sigma \propto (-\tilde{t})^\mu$, with $\mu = 3/2$ in the mean field approximation [see Eq. (65) in Sec. 2.7]. Likewise, the other exponent is connected with the divergence of the correlation length, ξ, which is the distance over which fluctuations of density or composition are strongly correlated and beyond which they are uncorrelated,

$\xi \propto (-\tilde{t})^{-\nu}$, with $\nu = 1/2$ in the mean field approximation [see Eq. (63) in Sec. 2.7]. Finally, there is another exponent, which is also associated with the correlations of fluctuations, but it does not concern us here.

These six exponents are not independent of each other, but are connected by the following scaling laws,

$$\delta = 1 + \gamma/\beta, \quad \mu + \nu = 2 - \alpha = \gamma + 2\beta, \tag{38}$$

and

$$d\nu = 2 - \alpha, \tag{39}$$

where d is the space dimensionality. It has been shown (Le Bellac, 1991, Ch. 1.3) that in four and higher dimensions the exponents given by the mean field theory are correct. In three dimensions, though, renormalization-group theory and experimental evidence have shown that $\alpha = 0.11$ (instead of 0), $\beta = 0.32$ (instead of 1/2), $\gamma = 1.24$ (instead of 1), $\delta = 4.8$ (instead of 3), $\mu = 1.26$ (instead of 3/2) and $\nu = 0.63$ (instead of 1/2). Despite these discrepancies between the behavior of a system near its critical point and the mean field predictions, the mean field theories, because of their simplicity, still have a prominent place in the statistical mechanics of multiphase systems (Widom, 1996). In addition, predictions based on the mean field theories compare favorably with molecular dynamics simulations, unless we consider systems that are very close to the critical point, where thermal fluctuations, which are neglected in the mean field theory, become very important. For example, in the vicinity of the vapor-liquid critical point, any fluctuations δv in the specific volumes of the two phases become comparable with the difference between their mean values, $\left(v_e^v - v_e^l\right)$, and therefore cannot be neglected. This is, in essence, the meaning of the Ginzburg criterion for the validity of the mean field theory, stating that $\langle(\delta\tilde{v})^2\rangle \ll \tilde{v}_e^2$. Clearly, this criterion strongly limits the region of applicability of the mean field theory, as we have imposed from the start that the system is close to its critical point so that, accordingly, its free energy can be expressed as a forth-order polynomial of the order parameter. On the other hand, the van der Waals theory is not confined to regions close to the critical point (actually, quite the contrary), and therefore this limitation is much less severe.

2.5 The diffuse interface

Suppose now that the molar density of the system is not constant. Accordingly, when $U_0 \ll kT$, Eq. (2) can be rewritten as

$$f(\mathbf{x}) = f_{Th}(\mathbf{x}) + \Delta f_{NL}(\mathbf{x}), \tag{40}$$

where f_{Th} is the molar free energy (3) corresponding to a system with constant density, while

$$\Delta f_{NL}(\mathbf{x}) = \frac{1}{2} N_A^2 \int_{r>d} U(r)[\rho(\mathbf{x}+\mathbf{r}) - \rho(\mathbf{x})] d^3\mathbf{r} \tag{41}$$

is a non local molar free energy, due to density changes, typical of the diffuse interface model. In fact, when there is an interface separating two phases at equilibrium, this term corresponds to the interfacial energy. This result is a direct consequence of the "exact" expression (2), showing that the free energy is non local, that is its value at any given point does not depend only on the density at that point, but it depends also on the density at neighboring points. As stated by van der Waals (1893), "the error that we commit in assuming a dependence on the density only at the point considered vanishes completely when the state of equilibrium is that of a homogeneous distribution of the substance. If, however, the state of equilibrium is one where there is a change of density throughout the vessel, as in a substance under the action of gravity, then the error becomes general, however feeble it may be." Now, in (41) the density can be expanded as

$$\rho(\mathbf{x}+\mathbf{r}) = \rho(\mathbf{x}) + \mathbf{r} \cdot \nabla \rho + \frac{1}{2} \mathbf{r}\mathbf{r} : \nabla\nabla\rho + \dots \tag{42}$$

As we have tacitly assumed that the system is isotropic, we see that the contribution of the linear term vanishes, so that, at leading order, we obtain (Pismen, 2001):

$$\Delta f_{NL}(\mathbf{x}) = -\frac{1}{2} RTK \nabla^2 \rho(\mathbf{x}), \tag{43}$$

with

$$K = \frac{2\pi}{3} \frac{N_A U_0}{kT} \frac{l^6}{d} = \frac{9\pi}{4} \frac{T_C}{T} N_A d^5, \tag{44}$$

where we have substituted Eqs. (4), (9) and (17). Note that, defining a non-dimensional molar density, $\tilde{\rho} = N_A d^3 \rho$, the non local free energy can be rewritten as

$$\Delta f_{NL}(\mathbf{x}) = -\frac{1}{2} RT a^2 \nabla^2 \tilde{\rho}(\mathbf{x}), \tag{45}$$

where

$$a = \sqrt{\frac{K}{N_A d^3}} = \sqrt{\frac{9\pi T_C}{4T}} d \tag{46}$$

is the characteristic length. Therefore, in the bulk, the total free energy is:

$$\int_V \hat{f} d^3\mathbf{x} = \int_V \rho \left(f_{Th} - \frac{1}{2} RTK\nabla^2 \rho \right) d^3\mathbf{x}, \tag{47}$$

where $\hat{f} = \rho f$ is the free energy per unit volume.

At the wall, the non local free energy (41) has an additional contribution of the form

$$\oint \left\{ \rho_s(\mathbf{x}) \left[\left[\int U(r) \rho(\mathbf{x}+\mathbf{r}) d^3\mathbf{r} \right] \right] \right\} d^2\mathbf{x} = \oint g_w(\rho(\mathbf{x})) d^2\mathbf{x}, \tag{48}$$

where the integration is carried out on the surface. Here, ρ_s is the solid density and g_w is the wall free energy per unit surface, that we assume to be a function of the fluid density at the wall only. Now, observing that, integrating by parts,

$$\int \rho(\mathbf{x}) \nabla^2 \rho(\mathbf{x}) d^3\mathbf{x} = \oint \mathbf{n} \cdot (\rho \nabla \rho) d^2\mathbf{x} - \int |\nabla \rho(\mathbf{x})|^2 d^3\mathbf{x}, \tag{49}$$

we see that the total free energy is the sum of a bulk and a surface free energies, i.e.,

$$F = F_b + F_w, \tag{50}$$

where F_b is the bulk free energy,

$$F_b = \int \hat{f}(\rho, \nabla\rho, T) d^3\mathbf{x} = RT \int \left[\rho \tilde{f}_{Th}(\rho, T) + \frac{1}{2} K(T)(\nabla\rho)^2 \right] d^3\mathbf{x}, \tag{51}$$

where $\tilde{f}_{Th} = f_{Th}/RT = \ln\rho + \rho B(T)$, while F_w is the wall free energy,

$$F_w = RT \oint \left[-\frac{1}{2} K\mathbf{n} \cdot (\rho\nabla\rho) + \tilde{g}_w(\rho) \right] d^2\mathbf{x}, \tag{52}$$

where $\tilde{g}_w = g_w/RT$.

At equilibrium, keeping the temperature T constant, the total free energy F will be minimized, subjected to the constraint of having a constant number of moles, $\int \rho d^3\mathbf{x} = const.$ Accordingly, introducing a Lagrange multiplier, $RT\tilde{\mu}$, the minimization condition is:

$$\delta \int \left[\rho \left(\tilde{f}_{Th}(\rho) - \tilde{\mu} \right) + \frac{1}{2} K |\nabla \rho|^2 \right] d^3\mathbf{x} + \delta \oint \left[-\frac{1}{2} RTK \mathbf{n} \cdot (\rho \nabla \rho) + g_w(\rho) \right] d^2\mathbf{x} = 0, \quad (53)$$

for any arbitrary variation $\delta\rho$ of the density field. Now, consider that, for any function $h(\rho, \nabla\rho)$, we have:

$$\delta h = \frac{\partial h}{\partial \rho} \delta\rho + \frac{\partial h}{\partial (\nabla_i \rho)} \delta(\nabla_i \rho), \quad \text{with } \delta(\nabla_i \rho) = \nabla_i (\delta \rho), \quad (54)$$

and

$$\int_V \frac{\partial h}{\partial \nabla_i \rho} \nabla_i (\delta \rho) d^3\mathbf{x} dt = \oint_S n_i \frac{\partial h}{\partial \nabla_i \rho} \delta \rho d^2\mathbf{x} - \int_V \nabla_i \left(\frac{\partial h}{\partial \nabla_i \rho} \right) \delta \rho d^3\mathbf{x}. \quad (55)$$

Applying these two equalities to Eq. (53) we obtain:

$$\int \left[\frac{\partial \hat{f}}{\partial \rho} - \nabla_i \left(\frac{\partial \hat{f}}{\partial \nabla_i \rho} \right) - (RT\tilde{\mu}) \right] \delta\rho d^3\mathbf{x} + \oint_S \mathbf{n} \cdot \left[\frac{1}{2} RTK \nabla \rho + \frac{d g_w}{d\rho} \right] \delta \rho d^3\mathbf{x} = 0, \quad (56)$$

where we have considered that $\delta(\rho\nabla\rho) = \delta\rho\,\nabla\rho + \rho\nabla\delta\rho$ and assumed that $\mathbf{n} \cdot \nabla\delta\rho = 0$ at the boundary.

2.6 The generalized chemical potential

Choosing $\delta\rho = 0$ at the boundary, Eq. (56) reduces to minimizing the bulk free energy. So, predictably, we obtain the Euler-Lagrange equation:

$$\tilde{\mu} = \frac{1}{RT} \frac{\delta(\rho f)}{\delta \rho} = \frac{d(\rho \tilde{f}_{Th})}{d\rho} - K \nabla^2 \rho. \quad (57)$$

Now, by definition, the first term on the RHS is the Gibbs free energy, which, in a one-component system, coincides with the chemical potential. In fact,

$$RT\tilde{\mu}_{Th} = \frac{d(\rho f_{Th})}{d\rho} = f_{Th} - v \frac{d f_{Th}}{dv}, \quad (58)$$

where $df_{Th}/dv = -P$. This (apart from the dimensional constant RT) is the equation of the straight line represented in Figure 2, stating that two phases at mutual equilibrium have the same chemical potential. Therefore, Eq. (57) can be rewritten as

$$\tilde{\mu}(\rho, \nabla\rho) = \tilde{\mu}_{Th}(\rho) - K\nabla^2\rho, \tag{59}$$

showing that at equilibrium, when ρ is non-uniform, it is $\tilde{\mu}$, and not $\tilde{\mu}_{Th}$, that remains uniform. Note that the thermodynamic chemical potential, $\tilde{\mu}_{Th}$, can be determined from the solvability condition of Eq. (58), that is,

$$\tilde{\mu}_{Th} = \frac{\rho^\alpha \tilde{f}^\alpha_{Th} - \rho^\beta \tilde{f}^\beta_{Th}}{\rho^\alpha - \rho^\beta} = \frac{v^\alpha \tilde{f}^\beta_{Th} - v^\beta \tilde{f}^\alpha_{Th}}{v^\alpha - v^\beta}, \tag{60}$$

as it can also be seen geometrically from Figure 2, stating that the chemical potential equals the intercept of the tangent line on the $v = 0$ vertical axis. When two phases are coexisting at equilibrium, separated by a planar interfacial region centered on $z = 0$, Eq. (57) can be solved once the equilibrium molar free energy f is known, imposing that, far from the interface region, the density is constant and equal to its equilibrium value, so that the generalized chemical potential is equal to its thermodynamic value (58). In particular, in the vicinity of the critical point, the chemical potential vanishes, the free energy is given by Eq. (32) and we obtain at leading order the following equation:

$$\frac{d^2\tilde{v}}{d\tilde{z}^2} - 2a_c\tilde{t}\tilde{v} - 4B_c\tilde{v}^3 = 0, \qquad \tilde{z} = z/\lambda, \qquad \lambda = \sqrt{\frac{K}{v_C(-a_c\tilde{t})}}, \tag{61}$$

to be solved imposing that

$$\tilde{v}(\tilde{z} \to \pm\infty) = \pm\tilde{v}_e = \pm\sqrt{\frac{-a_c\tilde{t}}{2B_c}}.$$

The solution, due (again!) to van der Waals, is:

$$\tilde{v}(\tilde{z}) = \tilde{v}_e \tanh \tilde{z}, \tag{62}$$

showing that λ is a typical interfacial thickness, basically coinciding with the correlation length mentioned in Section 2.4. For a van der Waals system, considering that K is given by Eq. (44), $v_C = 3c_2 = 2\pi d^3 N_A$ and $a_c = 3$, we obtain:

$$\lambda = \sqrt{\frac{a_c}{27(-\tilde{t})}\frac{l^3}{d^2}} = \sqrt{\frac{1}{8(-\tilde{t})}}d, \tag{63}$$

where we have substituted Eq. (19). As expected, the interfacial thickness diverges like $(-\tilde{t})^{-1/2}$ as we approach the critical point, while far from the critical point it is of $O(d)$. Recently, Pismen (2001) pointed out that Eq. (61) is flawed, as some of the neglected terms diverge at the critical point. In fact, Pismen showed that in the correct solution the specific volume tends to its equilibrium value as $|\tilde{z}|^{-3}$, instead of exponentially, as in the van der Waals solution.

2.7 The surface tension

In Section 2.5 we have seen that the total free energy is the sum of a thermodynamical, constant density, part, and a non local contribution (43). When the system is composed of two phases at equilibrium, separated by a plane interfacial region, we may define the surface tension as the energy per unit area stored in this region. This quantity can be calculated by integrating the specific free energy (43) along a coordinate z perpendicular to the interface:

$$\sigma = -\frac{1}{2}RTK\int_{-\infty}^{\infty}\rho\frac{d^2\rho}{dz^2}dz = \frac{1}{2}RTK\int_{-\infty}^{\infty}\left(\frac{d\rho}{dz}\right)^2 dz, \tag{64}$$

where we have integrated by parts and considered that, outside the interfacial region, the integrand is identically zero as density is constant. We see that, near the critical point,

$$\sigma \approx RT_C K(\Delta\rho_e)^2/\lambda \approx \frac{kT_C}{d^2}(-\tilde{t})^{3/2}, \tag{65}$$

where we have considered Eqs. (30), (44) and (63). In fact, using the density profile (62), Eq. (64) yields, for a van der Waals system:

$$\sigma = \frac{2}{3}RT_C(-a_c\tilde{t})^{3/2}\left(\frac{K}{v_c^3}\right) = C\frac{kT_C}{d^2}(-\tilde{t})^{3/2}, \tag{66}$$

with $C = 3^{3/2}/(2^{3/2}\pi)$, where we have used Eq. (18). These results show that the surface tension decreases as we approach criticality, until it vanishes at the critical point. A more detailed numerical solution based on the solution of the van der Waals equation can be found in Pismen and Pomeau (2000). Applying this approach, van der Waals (1893) showed that in a curved interface region there arises a net force [see Section 4.1], which is compensated by a pressure term, thus obtaining the Young-Laplace equation. To see that, let us denote the position of the interface by $z = h(\xi)$, where ξ is the 2D vector in the support plane, and assume that $|\nabla_\xi h| \ll 1$, where $|\nabla_\xi h|$ is the 2D gradient (Pismen, 2001). Now, the free energy increment due to the interface curvature can be written as:

$$\Delta F = \sigma \int \left(\sqrt{1 + |\nabla_\xi h|^2} - 1 \right) d^2\xi \approx \frac{1}{2}\sigma \int |\nabla_\xi h|^2 d^2\xi. \tag{67}$$

This increment in the free energy induces an increment in the pressure,

$$\Delta P = \delta F/\delta h = -\sigma \nabla^2 h = -\kappa\sigma, \tag{68}$$

where $\kappa = \nabla^2 h$ is the curvature of a weakly curved interface. Applying a rigorous regular perturbation approach to Eq. (57), Pismen and Pomeau (2000) derived both the Young-Laplace equation (68) and the Gibbs-Thomson law, relating the equilibrium temperature or pressure to the interfacial curvature.

The mean field prediction (66) is not too different than the "exact" result, obtained by renormalization group theory, showing that surface tension vanishes as temperature approaches its critical value with a 1.26 (instead of 1.5) critical exponent.

2.8 The boundary conditions

Choosing $\delta\rho = 0$ in the bulk, Eq. (56) reduces to minimizing the surface integral, imposing local equilibrium at the wall. Therefore, we obtain the following boundary condition (Jacqmin, 2000),

$$\frac{1}{2} RTK\mathbf{n} \cdot \nabla\rho = -\frac{dg_w}{d\rho}(\rho). \tag{69}$$

It is worth noting that this is an equilibrium boundary condition. However, this condition can be extended to out-of-equilibrium conditions to recover a variation of the contact angle with the speed of displacement of the contact line (a variation that is observed experimentally).

3 Binary mixtures

In this Section we will show that van der Waals' approach can be applied to study binary solutions as well. To simplify matters, at first let us confine ourselves to the case of regular binary solutions, that are mixtures in which both the excess volume of mixing and the excess entropy of mixing are equal to zero. That means that when we mix the two species, say 1 and 2, a) the volume remains unchanged, so that the mixture can be considered to be incompressible, and b) the entropy change is equal to that of ideal mixtures (Sandler, 1999, Ch. 7.6). Generalization to non regular, even compressible, binary mixtures can be found in Lowengrub and Truskinovsky (1998).

3.1 The Gibbs free energy

The molar free energy can be determined using the same procedure as for single component systems. Consider a mixture composed of species 1 and species 2, with molar fraction $x_1 = \phi$ and $x_2 = (1 - \phi)$ and let us first determine the molar free energy when the composition of the mixture is uniform. Considering the definition of Gibbs free energy, $g = f + P/\rho$, from Eq. (2) we obtain

$$g_{Th}(\phi) = g_{id}(\phi) + g_{ex}(\phi). \tag{70}$$

Here g_{id} is the Gibbs free energy of an ideal mixture, that is a mixture where the intermolecular potentials U_{ij} between molecule i and molecule j are all the same, i.e. $U_{11} = U_{22} = U_{12}$. Generalizing the expression of the free energy for a single component fluid, $RT \ln \rho$, we obtain:

$$g_{id} = RT[x_1 \ln(x_1 \rho) + x_2 \ln(x_2 \rho)],$$

that is

$$g_{id} = RT \ln \rho + RT[\phi \log \phi + (1 - \phi) \log(1 - \phi)]. \tag{71}$$

Note that for a pure fluid the molar density ρ is a variable, while for a regular binary mixture the total molar density ρ can even be constant, since the variables are the molar densities of the two components, $x_1 \rho$ and $x_2 \rho$.

The second term in the RHS of Eq. (70), g_{ex}, is the so called excess, that is non ideal, part of the free energy. This term has a particularly convenient form for regular mixtures, as it is explained below. The theory of regular mixtures was developed by van Laar (a student of van der Waals), who assumed that (a) the two species composing the mixture are of similar size and energies of interaction, and (b) the van der Waals equation of state applies to both the pure fluids and the mixture. Consequently, regular mixtures have negligible excess volume and excess entropy of mixing, i.e. their volume and entropy coincide with those of an ideal gas mixture, with $s_{ex} = 0$ and $v_{ex} = 0$. Therefore, we see that for a regular mixture, since $s_{ex} = -(\partial g_{ex}/\partial T)_{P,x} = 0$, then g_{ex} must be independent of T. In addition, as the excess Gibbs free energy results to be equal to the excess internal energy, i.e. $g_{ex} = u_{ex}$, it can be shown (Sandler, 1999, Ch. 7.6) that the Gibbs free energy of a regular mixture with molar fractions x_1 and $x_2 = 1 - x_1$ is:

$$g_{ex} = x_1 \frac{c_1^{(1)}}{c_2^{(1)}} + x_2 \frac{c_1^{(2)}}{c_2^{(2)}} - \frac{c_1^{mix}}{c_2^{mix}}, \tag{72}$$

where we have used obvious notations to indicate the van der Waals constants (c_1 and c_2) for the pure fluids, 1 and 2, and for the mixture, and we have considered that $u_{ex} = c_1/c_2$. From here, assuming that

$$c_2^{mix} = c_2^{(1)} = c_2^{(2)}, \quad c_1^{mix} = x_1^2 c_1^{(1)} + x_2^2 c_1^{(2)} + 2x_1 x_2 c_1^{(12)}, \tag{73}$$

with $c_1^{(12)} = \sqrt{c_1^{(1)} c_1^{(2)}}$, we obtain,

$$g_{ex} = \frac{1}{c_2} x_1 x_2 (c_1^{(1)} + c_1^{(2)} - 2\sqrt{c_1^{(1)} c_1^{(2)}}). \tag{74}$$

Thus, we see again that the excess free energy does not depend on T, so that $s_{ex} = 0$. Similar considerations can be applied to the dependence of the free energy on the pressure, confirming that $v_{ex} = 0$ for regular mixtures.

The same conclusions can be reached starting from the fundamental expression (2) for the Helmholtz free energy and considering that:

$$g_{ex} = f_{ex} + P v_{ex}. \tag{75}$$

Applying Eqs. (6), (7) and (75) to a system with constant molar density ρ, i.e. with $v_{ex} = 0$, we obtain $g_{ex} = f_{ex} = RT\rho B$, where B is the virial coefficient:

$$B = x_1^2 B^{(11)} + 2x_1 x_2 B^{(12)} + x_2^2 B^{(22)}. \tag{76}$$

Here, $B^{(ij)}$ characterizes the repulsive interaction between molecule i and molecule j [see Eq. (7)],

$$B^{(ij)} = \frac{1}{2} N_A \int \left[1 - \exp\left(-\frac{U^{(ij)}(r)}{kT}\right)\right] d^3\mathbf{r}, \tag{77}$$

where $U^{(ij)}$ is the pairwise interaction potential between molecules i and j. In particular, for symmetric solutions, $U^{(11)} = U^{(22)} \neq U^{(12)}$, so that $B^{(11)} = B^{(22)} \neq B^{(12)}$. Accordingly, denoting $x_1 = \phi$, we obtain:

$$g_{ex} = \rho RTB = 2\rho RT(B^{(12)} - B^{(11)})\phi(1-\phi),$$

that is

$$g_{ex}(T, P, \phi) = RT\Psi(T, P)\phi(1-\phi), \tag{78}$$

where

$$\Psi(T, P) = 2\rho(B^{(12)} - B^{(11)}), \tag{79}$$

is the so called Margules coefficient (Sandler, 1999). In particular, for an ideal mixture, $B^{(11)} = B^{(12)}$ and therefore $\Psi = 0$. For a mixture composed of van der Waals fluids at constant pressure, substituting the expression (8) for B and assuming that the characteristic lengths d and l are the same for the two species, we obtain:

$$\Psi = \frac{2\rho}{RT}(c_1^{(11)} - c_1^{(12)}) = \frac{4\pi}{3}\frac{\rho N_A^2 l^6}{RTd^3}(U_0^{(11)} - U_0^{(12)}), \tag{80}$$

where $U_0^{(11)}$ and $U_0^{(12)}$ characterize the strength of the potential between molecules of the same species and that of different species, respectively. From this expression we see that $\Psi \propto T^{-1}$, confirming that g_{ex} is independent of T. Note that when $\Psi > 0$ the repulsive forces $\cong U_0^{(12)}/d$ between unlike molecules are weaker than those between like molecules, $\cong U_0^{(11)}/d$. As shown in Mauri et al. (1996), when the solution is not symmetric, this approach is easily generalized by defining two Margules coefficients.

3.2 Coexistence and spinodal curve

In the previous Section, we saw that the free energy of a homogenous symmetric binary mixture can be written as:

$$g_{Th} = g_1 + RT[\phi \log \phi + (1-\phi)\log(1-\phi) + \Psi\phi(1-\phi)]. \tag{81}$$

Now, the thermodynamic state of a one-component system is determined by fixing two quantities, e.g. P and T. In binary mixtures, we have an additional degree of freedom, i.e. the molar fraction of the two species, x_1. Associated with x_1 and x_2 we can define the respective chemical potentials, $RT\mu_i = \partial(Ng_{Th}/\partial N_i)_{N_{j\neq i}}$. Generalizing the relation obtained in the previous Section, at equilibrium the chemical potentials μ_1 and μ_2 are uniform everywhere and, in particular, they must be the same in each phase, i.e., $\mu_1^\alpha = \mu_1^\beta$ and $\mu_2^\alpha = \mu_2^\beta$. Considering that x_1 and x_2 depend on each other, there is a relation between μ_1 and μ_2, namely the Gibbs-Duhem relation (Sandler, 1999), $x_1\nabla\mu_1 = -x_2\nabla\mu_2$. This relation can be easily obtained by imposing that the specific Gibbs free energy, g_{th}, defined as

$$Ng_{Th} = Nu_{Th} - NTs + NPv + RTN_1\mu_1 + RTN_2\mu_2, \tag{82}$$

where u_{Th} is the molar internal energy, must satisfy the following equality,

$$dg_{Th} = -sdT + vdP + RT\mu d\phi, \tag{83}$$

where $\mu = \mu_1 - \mu_2$ is the chemical potential difference. This last relation reveals that the chemical potential difference is the quantity which is thermodynamically conjugated with the composition ϕ. This same result can be obtained from the identities (Sandler, 1999),

$$RT\mu_1(T,P,\phi) = g_{Th}(T,P,\phi) + \left(\frac{dg_{Th}}{d\phi}\right)(1-\phi), \tag{84}$$

$$RT\mu_2(T,P,\phi) = g_{Th}(T,P,\phi) - \left(\frac{dg_{Th}}{d\phi}\right)\phi, \tag{85}$$

obtaining

$$\mu = \mu_1 - \mu_2 = \frac{d(g_{Th}/RT)}{d\phi} = \log\left(\frac{\phi}{1-\phi}\right) + \Psi(1-2\phi), \tag{86}$$

where we have substituted Eq. (81). At constant temperature T and pressure P, since Ψ is a known function of T and P, this equation gives the dependence of the chemical potential difference on the composition, just like the equation of state, e.g. van der Waals' equation, gives the dependence of the pressure on the specific volume. Clearly, μ represents the tangent to the free energy curve and it is the same for the two phases at equilibrium. Accordingly, this equation leads to the determination of the equilibrium composition of the two coexisting phases, ϕ_e^α and ϕ_e^β. In particular, in our case, where we have considered symmetric mixtures, the tangent is horizontal and therefore $\mu = 0$. Note that, as expected, in this case $\phi_e^\alpha = 1 - \phi_e^\beta$. Now we can apply to binary mixtures the same considerations about the critical point that we made in the previous Section on one-component systems, observing that chemical potential difference μ and composition ϕ play the same role as pressure and density (or specific volume). In fact, in the $\mu - T$ diagram, the liquid-liquid equilibrium curve stops at the critical point, characterized by a critical temperature T_C and a critical chemical potential difference, $\mu_C = 0$ (note that for symmetric mixtures $\mu = 0$ at any equilibrium state, while in general that is true only at the critical point). At higher temperatures, $T > T_C$, the differences between the two liquid phases vanish altogether and the system is always in a single phase. In addition, as the critical point is approached from below (i.e. with two coexisting phases), the difference between the composition of the two phases decreases, until it vanishes altogether at the critical point. Accordingly, near the critical point, since the composition of the two phases, ϕ and $\phi + \delta\phi$, are near to each other, we obtain (Landau and Lifshitz, 1980, Ch. 97):

$$0 = \mu_C(T,\phi) = \mu_C(T,\phi+\delta\phi) \quad \Longrightarrow \quad 0 = \left(\frac{\partial\mu}{\partial\phi}\right)_C \delta\phi + \frac{1}{2}\left(\frac{\partial^2\mu}{\partial\phi^2}\right)_C (\delta\phi)^2 + \ldots,$$

where we have considered that, since the two phases are at equilibrium, they have the same chemical potential (in addition to having the same pressure and temperature). At this point, dividing by $\delta\phi$ and letting $\delta\phi \to 0$, we see that at the critical point we have:

$$\left(\frac{\partial \mu}{\partial \phi}\right)_{T_C,P_C} = 0. \tag{87}$$

Note that this condition is the limit case of the inequality $(\partial\mu/\partial\phi)_{T,P} \geq 0$, which manifests the internal stability of any two-phase system. In addition, expanding δg_{Th} in a power series of $\delta\phi$, with constant T and P, we obtain:

$$\delta g_{Th} = \left(\frac{\partial g_{Th}}{\partial \phi}\right)_{T,P}(\delta\phi)+\frac{1}{2!}\left(\frac{\partial^2 g_{Th}}{\partial \phi^2}\right)_{T,P}(\delta\phi)^2+\frac{1}{3!}\left(\frac{\partial^3 g_{Th}}{\partial \phi^3}\right)_{T,P}(\delta\phi)^3+\ldots$$

Therefore, since near an equilibrium point we have:

$$\delta u_{Th} - T\delta s + P\delta v - RT\mu\delta\phi = \delta g_{Th} - RT\mu\delta\phi > 0, \tag{88}$$

considering that $(\partial g_{Th}/\partial\phi)_{T,P} = RT\mu$ and that $(\partial^2 g_{Th}/\partial\phi^2)_{T,P} = 0$ at the critical point, we obtain:

$$\frac{1}{3!}\left(\frac{\partial^2 \mu}{\partial \phi^2}\right)_C(\delta\phi)^3 + \frac{1}{4!}\left(\frac{\partial^3 \mu}{\partial \phi^3}\right)_C(\delta\phi)^4 + \ldots > 0.$$

Since this equality must be valid for any value (albeit small) of $\delta\phi$ (both positive and negative), we obtain:

$$\left(\frac{\partial^2 \mu}{\partial \phi^2}\right)_{T_C,P_C} = 0, \qquad \left(\frac{\partial^3 \mu}{\partial \phi^3}\right)_{T_C,P_C} > 0. \tag{89}$$

Therefore, the critical point corresponds to a horizontal inflection point in the $\mu - \phi$ diagram (see Figure 1), which means that,

$$\left(\frac{\partial^2 g_{Th}}{\partial \phi^2}\right)_{T_C,P_C} = 0, \qquad \left(\frac{\partial^3 g_{Th}}{\partial \phi^3}\right)_{T_C,P_C} = 0. \tag{90}$$

Imposing that at the critical point the $\mu - \phi$ curve has a horizontal inflection point, from Eq. (86) we see that $\phi_C^\alpha = \phi_C^\beta = 1/2$, confirming that $\Psi_C = 2$. Therefore, considering that $\Psi \propto T^{-1}$, we obtain:

$$\Psi = \frac{2T_C}{T}. \tag{91}$$

In particular, near the critical point, defining $\tilde{\psi} = (\Psi - 2)/2$ and $\tilde{t} = (T - T_C)/T_C$, we obtain at leading order:

$$\tilde{\psi} = -\tilde{t}. \tag{92}$$

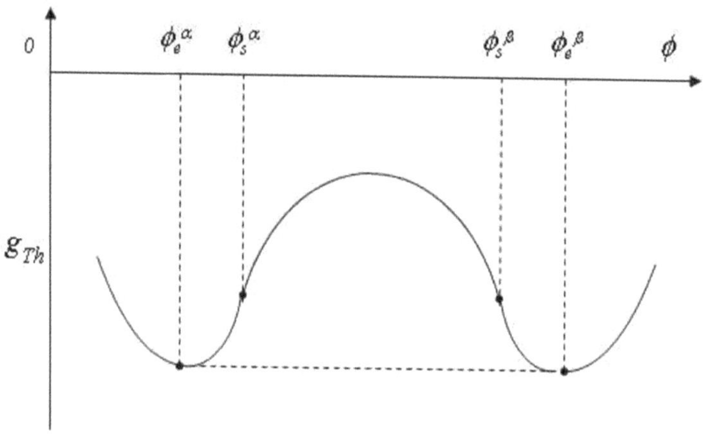

Figure 4. Typical double-well curve of the free energy of a symmetric binary mixture.

Now, let us consider a binary liquid mixture at equilibrium, whose chemical potential difference and temperature are below their critical values, so that the mixture is separated into two coexisting phases. At equilibrium, the phase transition takes place at constant temperature, pressure and chemical potential difference and therefore it can be represented as a horizontal isotherm isobaric segment in a $\mu - \phi$ diagram. Now, define a generalized potential (Landau and Lifshitz, 1980, Ch. 85) as $\Phi_{Th} = g_{Th} - \mu\phi$, with $d\Phi_{Th} = -sdT + vdP - RT\phi d\mu$ and $(\partial\Phi/\partial\mu)_{T,P} = -RT\phi$. The chemical potential difference μ at a given temperature and pressure can be easily determined, considering that at equilibrium the generalized potentials of the two phases must be equal to each other, that is

$$\Phi_{Th}^{\beta} - \Phi_{Th}^{\alpha} = \int_b^e d\Phi_{Th} = 0 \quad \Longrightarrow \quad \int_b^e \phi d\mu = [\mu\phi]_b^e - \int_b^e \mu d\phi = 0, \tag{93}$$

where we have considered that the phase transition is isothermal and isobaric. From a geometrical point of view, this relation manifests the equality between the shaded area of Figure 1 (Maxwell's rule), where the points b and e correspond to the saturation points of the two phases at that temperature and pressure, with compositions ϕ^α and ϕ^β. Conversely, the composition of the two phases at equilibrium can be determined from the molar free energy g of Eq. (81) as follows. A typical curve of the free energy g for a symmetric binary mixture is represented in Figure 4, when $T < T_C$ (i.e. when $\Psi > 2$). The condition (93) expresses the fact that at equilibrium the two phases have the same temperature, pressure and chemical potential (so that the chemical potential difference μ is identically zero). Being on an isotherm-isobar, the first two conditions are automatically satisfied while, using the relation $\mu = (\partial g/\partial \phi)_{T,P}$, the last condition gives:

$$\mu^\alpha = \mu^\beta = 0 \quad \Longrightarrow \quad \left(\frac{\partial g_{Th}}{\partial \phi}\right)^\alpha_{T,P} = \left(\frac{\partial g_{Th}}{\partial \phi}\right)^\beta_{T,P} \quad \Longrightarrow \quad (\phi^\alpha_e, \phi^\beta_e), \qquad (94)$$

which, in Figure 4, represents the fact that the two equilibrium points have the same tangent and this tangent is a horizontal line. When the mixture is not symmetric, the $g_{Th} - \phi$ curve is similar to the $f_{Th} - v$ curve of Figure 2. Consequently, it is still true that $\mu^\alpha = \mu^\beta$, but, in general, they are not equal to zero, i.e. the tangent to the Gibbs free energy curve is not horizontal. In Figure 1, besides the equilibrium curve, we have represented the, so called, spinodal curve, defined as the locus of all points (like c and d) satisfying $(\partial \mu/\partial \phi)_{T,P} = 0$. When the equilibrium and spinodal points are plotted in a $T - \phi$ diagram (at constant pressure), we obtain the curves of Figure 3. All points lying outside the region encompassing the equilibrium curve represent homogeneous, single-phase mixtures in a state of stable equilibrium, while all points lying inside that region represent systems in a state of non equilibrium, which tend to separate into two phases. However, all points lying in the region inside the spinodal curve are unstable, that is any infinitesimal perturbation can trigger the phase transition process, while all points sandwiched between the equilibrium and the spinodal curves represent metastable systems, i.e. mixtures that need an activation energy to phase separate. The spinodal points can be also determined using the relation $(\partial \mu/\partial \psi)_{T,P} = (\partial^2 g_{Th}/\partial \phi^2)_{T,P} = 0$, obtaining:

$$\left(\frac{\partial^2 g_{Th}}{\partial \phi^2}\right)_{T,P} = 0 \quad \Longrightarrow \quad (\phi^\alpha_s, \phi^\beta_s). \qquad (95)$$

3.3 The critical exponents

Let us turn now to study the behavior of a binary mixture close to its critical point. Then, instead of Ψ and ϕ, it is convenient to use the following variables,

$$\tilde{\psi} = \frac{1}{2}(\Psi - 2), \quad \tilde{u} = 2\phi - 1. \tag{96}$$

Consequently, neglecting higher order terms, we obtain from the equation of state (86),

$$\mu = -2\tilde{\psi}\tilde{u} + \frac{2}{3}\tilde{u}^3. \tag{97}$$

Note that we cannot have any term proportional to $\tilde{u}$ or $\tilde{u}^2$, in agreement with the conditions $(\partial\mu/\partial\phi)_{T_c,P_c} = 0$ and $(\partial^2\mu/\partial\phi^2)_{T_c,P_c} = 0$, while the coefficient of the $\tilde{u}^3$-term must be positive, as $(\partial^3\mu/\partial\phi^3)_{T_c,P_c} > 0$. When $\tilde{\psi} < 0$, all states are stable, that is there is no phase separation and the mixture remains homogeneous. This means that, when $\tilde{\psi} < 0$, it must be $(\partial\mu/\partial\phi)_{T,P} > 0$, and therefore the coefficient of the term $\tilde{\psi}\tilde{u}$ must be negative. Equation (97) is a particular case of the general expression (28), with $a_c = 1, b_c = 0$ and $B_c = -1/6$. Accordingly, all considerations that we made in Section 2.4 relative to this expression can be repeated now. In particular, the critical exponents are the same. Based on the expression (97) it is easy to evaluate the composition of the two coexisting phases at equilibrium and those on the spinodal curve near the critical point. In fact, at constant temperature and pressure (and therefore at constant Ψ) we saw that $(\partial g/\partial\phi)_\Psi = 0$, so that $\mu = \mu_1 - \mu_2 = 0$. Therefore, from the expansion (97), we obtain, near the critical point:

$$\tilde{u}_e = \pm\sqrt{3\tilde{\psi}}. \tag{98}$$

The spinodal point, instead, can be determined via Eq. (95), obtaining, near the critical point:

$$\tilde{u}_s = \pm\sqrt{\tilde{\psi}}. \tag{99}$$

Note that, as expected, $|\tilde{u}_s| < |\tilde{u}_e|$.

3.4 The diffuse interface

Suppose now that the composition of the system is not constant. Accordingly, we have:

$$g(\mathbf{x}) = g_{Th}(\mathbf{x}) + \Delta g_{NL}(\mathbf{x}), \tag{100}$$

where g_{Th} is the molar free energy (81) corresponding to a system with constant density,

$$g_{Th} = [g_1\phi + g_2(1-\phi)] + RT[\phi\log\phi + (1-\phi)\log(1-\phi) + \Psi\phi(1-\phi)], \tag{101}$$

while Δg_{NL} is a non local molar free energy, due to changes in composition, typical of the diffuse interface model. In fact, when there is an interface separating two phases at equilibrium, this term corresponds to the interfacial energy. Using a procedure similar to that seen in the previous Section, we can derive an expression, originally due to Cahn and Hilliard (1958),

$$\Delta g_{NL} = \Delta g_{CH} = \frac{1}{2}RTa^2(\nabla\phi)^2, \tag{102}$$

stating that whenever there is an interface, or even a change in composition, there must be an increase of energy (see van der Waals, 1893, for an extended discussion about this term). Therefore, we may conclude that the expression for the bulk free energy in this case is basically identical to that of a single-phase fluid [cf. Eq. (51)], i.e.,

$$g(\phi, \nabla\phi) = g_{Th}(\phi) + \frac{1}{2}RTa^2(\nabla\phi)^2. \tag{103}$$

Here, a is a characteristic length, roughly equal to the interface thickness at equilibrium which, for a regular mixture, has the same value as that seen in Eq. (46), i.e.,

$$a = \sqrt{\frac{9\pi T_C}{4T}}d, \tag{104}$$

where d is the excluded volume length defined in (9). Now, following the same procedure as in Section 2.7, observe that at the end of the phase segregation process, a surface tension σ can be measured at the interface and from that, as shown by van der Waals (1893), a can be determined as

$$a \approx \frac{1}{\sqrt{\tilde{\psi}(\Delta\phi)^2_{eq}}}\frac{\sigma M_w}{\rho RT}, \tag{105}$$

where $\tilde{\psi} = (\Psi - 2)/2$, while $(\Delta\phi)_{eq}$ is the composition difference between the two phases at equilibrium. This relation can be easily derived considering that at equilibrium the surface tension σ is equal to the integral of the

Cahn-Hilliard free energy across the interface, i.e. $\sigma \approx (\rho l/M_w)\Delta g_{eq}$, where $\Delta g_{eq} \approx RT(\Delta\phi_{eq})^2 a^2/l^2$ is a typical value of the change in the Cahn-Hilliard molar free energy within the interface at equilibrium, while $l \approx a/\sqrt{\tilde{\psi}}$ is the characteristic interface thickness (van der Waals, 1893).

3.5 The generalized chemical potential

At equilibrium, the total free energy is minimized. As we saw in the previous Section, minimization can be carried out separately in the bulk and at the boundaries. Accordingly, in the bulk we have:

$$\int_V g(\phi, \nabla\phi) d^3\mathbf{x} = min., \tag{106}$$

with the constraint

$$\int_V \phi(\mathbf{x}) d^3\mathbf{x} = const. \tag{107}$$

Therefore, applying to the system a virtual change in composition, $\delta\phi(\mathbf{x})$, and proceeding as in the previous Section, we obtain, as expected, the Euler-Lagrange equation,

$$\tilde{\mu} = \frac{1}{RT}\frac{\delta g}{\delta\phi} = \frac{1}{RT}\left(\frac{\partial g}{\partial\phi} - \nabla_i \frac{\partial g}{\partial(\nabla_i\phi)}\right) = \mu(\phi) - \frac{1}{RT}\nabla_i \frac{\partial g}{\partial(\nabla_i\phi)}, \tag{108}$$

showing that $\tilde{\mu}$ is the generalized chemical potential difference, which must be uniform at equilibrium. Substituting Eq. (103) and reminding that $\mu = \mu_1 - \mu_2$ is the difference between the thermodynamical chemical potentials of the two species (92), we obtain:

$$\tilde{\mu}(\phi, \nabla\phi) = \mu(\phi) - a^2\nabla^2\phi. \tag{109}$$

Near the critical point, the generalized chemical potential difference is zero and therefore, in 1D, along z, substituting Eq. (97) into (109), we obtain:

$$\partial_{zz}\tilde{u} + 4\tilde{\psi}\tilde{u} - (4/3)\tilde{u}^3 = 0, \tag{110}$$

with $\partial_z = \partial/\partial\tilde{z}$, where $O(\tilde{u}^5)$ terms have been neglected, while $\tilde{u}$ and $\tilde{\psi}$ have been defined in (96). Solving this equation in an unbounded domain, imposing that $\tilde{u}(\pm\infty) = \pm\tilde{u}_e = \pm\sqrt{3\tilde{\psi}}$ (this is the composition at equilibrium, when $\tilde{\psi} \ll 1$) we obtain van der Waals' solution,

$$\tilde{u}(z) = \tilde{u}_e \tanh(\sqrt{2\tilde{\psi}}z/a). \tag{111}$$

As shown in Mauri et al. (1996), this solution can be generalized to finite systems, obtaining a family of Jacobi's elliptic functions. This solution shows that the typical interface thickness l is of $O(a/\sqrt{\tilde{\psi}})$. As we have remarked in the previous Section, Pismen (2001) solved the full problem, showing that $\tilde{u}$ tends to its equilibrium value as $|\tilde{z}|^{-4}$, instead of exponentially, as in the van der Waals solution.

3.6 The boundary conditions

As previously noted, the equilibrium state of an unconfined van der Waals fluid can be determined using the generalized chemical potential in the bulk. In general, however, for confined systems surface wettability effects are present and must be taken into account. In our D.I. approach such effects have been accounted for by introducing the simplest additional surface contribution to the free energy functional, which is based on the assumption that wettability is a local quantity, depending on the composition of the mixture at the wall, i.e.,

$$\int g_w(\phi)\, d^2\mathbf{x}. \tag{112}$$

The resulting wall boundary condition describes a diffusively controlled local equilibrium at the wall.

Proceeding in the same way as in Section 2.8, we obtain the following equilibrium boundary condition:

$$\frac{1}{2}a^2\rho RT\mathbf{n}\cdot\nabla\phi = -\frac{dg_w}{d\phi}(\phi), \tag{113}$$

where $g_w(\phi)$ is the surface energy at the wall. Assuming a linear dependence,

$$g_w(\phi) = \phi\sigma_{Aw} + (1-\phi)\sigma_{Bw}, \tag{114}$$

and considering that $\sigma \cong a\rho RT/M_w$ is the surface tension between the two fluid phases at equilibrium, this condition can be rewritten as

$$\sigma\, a\, \mathbf{n}\cdot\nabla\phi = -\Delta\sigma_w, \tag{115}$$

where $\Delta\sigma_w = \sigma_{Aw} - \sigma_{Bw}$ expresses the affinity of the wall to species A, as compared to species B. This condition is generally referred to as the Cahn (1977) boundary condition. In the sharp interface limit, $\mathbf{n}\cdot\nabla\phi = \cos\theta$, where θ is the contact angle, and therefore the Cahn boundary condition reduces to the Young-Laplace formula, $\cos\theta = -\Delta\sigma_w/\sigma$. From here we see that, when $\sigma_{Aw} = \sigma_{Bw}$ or when σ_{Aw} and σ_{Bw} are both $\ll \sigma$, then $\theta = \pi/2$; instead, when $\sigma_{Aw} \gg \sigma_{Bw}$ or $\sigma_{Aw} \ll \sigma_{Bw}$, then $\theta = \pi$ and $\theta = 0$, respectively.

4 Equations of motion for non-dissipative mixtures

4.1 The Korteweg stresses

In this Section, we confine ourselves to study a binary mixture with constant density, composed of two species with the same molecular weight, M_w. However, the case of a single-component system and that of non-symmetric and even compressible binary mixtures can be handled in the same way, obtaining very similar results. These generalizations can be found in Lowengrub and Truskinovsky (1998), Anderson et al. (1998) and, more recently, in Thiele et al. (2007) and Madruga and Thiele (2009). First, let us consider the reversible, dissipation-free case. Then, there is no diffusion and therefore the concentration field can be derived from the initial conditions, knowing the velocity field. In fact, if $\mathbf{x}(t, \mathbf{x}_0)$ denotes the trajectory of a material particle which is located at $\mathbf{x}_0$ at time $t = 0$, i.e. with $\mathbf{x}_0 = \mathbf{x}(0, \mathbf{x}_0)$, then the fluid velocity field is $\mathbf{v}(\mathbf{x}, t) = \dot{\mathbf{x}}(t, \mathbf{x}_0)$, where the dot denotes time derivative at constant $\mathbf{x}_0$, while the concentration field $\phi(\mathbf{x}, t) = \phi(\mathbf{x}_0)$ does not depend explicitly on time, and therefore $\dot{\phi} = 0$. According to the Hamilton, minimum action, principle, the motion of any conservative system minimizes the following functional:

$$S = \int_0^t \int_V \mathscr{L}(\mathbf{v}, \phi, \nabla\phi) d^3\mathbf{x} dt, \tag{116}$$

where

$$\mathscr{L} = \mathscr{L}(\mathbf{v}, \phi, \nabla\phi) = \rho\left(\frac{1}{2}v^2 - \frac{1}{M_w}g\right) \tag{117}$$

is the Lagrangian of the system, subjected to the constraint of incompressibility,

$$\nabla \cdot \mathbf{v} = 0. \tag{118}$$

Accordingly, starting from the minimum condition, let us give a virtual displacement δx_i, corresponding to an infinitesimal change of the fluid flow. Among all the possible virtual displacements, let us choose those such that $\delta\phi = \nabla_i\phi \cdot \delta x_i = 0$. Since the action S in (116) is minimized, we have:

$$\delta S = \int_0^t \int_V \left[\rho v_i \delta v_i - \frac{\rho}{M_w}\delta g - q\delta(\nabla_i v_i)\right] d^3\mathbf{x} dt = 0, \tag{119}$$

where $q(\mathbf{x}, t)$ is a function to be determined through the constraint (118). Note that the constraint (107) does not apply here because the concentration field remains unchanged, following the evolution equation $\dot{\phi} = 0$. Considering that $\delta(dx_i) = d(\delta x_i)$, the first integral term on the RHS of Eq. (119) gives, after integrating by parts:

$$\int_0^t \int_V \rho v_i \delta v_i d^3\mathbf{x}dt = \int_0^t \int_V \rho v_i \delta \frac{dx_i}{dt} d^3\mathbf{x}dt$$
$$= \int_0^t \int_V \rho v_i \frac{d}{dt}(\delta x_i) d^3\mathbf{x} = \int_V [\rho v_i \delta x_i]_{t_1}^{t_2} d^3\mathbf{x} - \int_0^t \int_V \rho \frac{dv_i}{dt} \delta x_i d^3\mathbf{x}dt, \quad (120)$$

i.e.

$$\int_0^t \int_V \rho v_i \delta v_i d^3\mathbf{x}dt = -\int_0^t \int_V \rho \frac{dv_i}{dt} \delta x_i d^3\mathbf{x}dt. \quad (121)$$

Here we have considered that the virtual displacement is equal to zero at the beginning and at the end, i.e. when $t = t_1$ and $t = t_2$, as well as on the boundary, S, of the volume V of integration.

The second integral term on the RHS of Eq. (119) gives, considering that $g = g(\phi, \nabla\phi)$:

$$\int_0^t \int_V \delta g d^3\mathbf{x}dt = \int_0^t \int_V \frac{\partial g}{\partial \phi} \delta\phi d^3\mathbf{x}dt + \int_0^t \int_V \frac{\partial g}{\partial \nabla_i \phi} \delta(\nabla_i \phi) d^3\mathbf{x}dt. \quad (122)$$

Now, considering that $\delta\phi = 0$, the first term on the RHS is identically zero. In addition, applying the following equality,

$$\delta(\nabla_j \phi) = \frac{\partial(\nabla_j \phi)}{\partial x_i} \delta x_i = (\nabla_i \nabla_j \phi)\delta x_i = \nabla_j(\delta\phi) - (\nabla_i \phi)\nabla_j(\delta x_i) = -(\nabla_i \phi)\nabla_j(\delta x_i),$$

we obtain:

$$\int_0^t \int_V \delta g d^3\mathbf{x}dt = -\int_0^t \int_V \frac{\partial g}{\partial \nabla_j \phi} (\nabla_i \phi)(\nabla_j \delta x_i) d^3\mathbf{x}dt$$
$$= -\int_0^t \oint_S n_j \frac{\partial g}{\partial \nabla_j \phi} \nabla_i \phi \delta x_i dS dt + \int_0^t \int_V \nabla_j \left(\frac{\partial g}{\partial(\nabla_j \phi)} \nabla_i \phi \right) \delta x_i d^3\mathbf{x}dt, \quad (123)$$

where we have integrated by parts. Here, too, the surface integral is identically zero, because $\delta x_i = 0$ at the boundary, so that

$$\int_0^t \int_V \delta g d^3\mathbf{x}dt = \int_0^t \int_V \nabla_j \left(\frac{\partial g}{\partial(\nabla_j \phi)} \nabla_i \phi \right) \delta x_i d^3\mathbf{x}dt. \quad (124)$$

Finally, the last integral term on the RHS of (119) gives:

$$\begin{aligned}\int_0^t \int_V q\delta(\nabla_i v_i) d^3\mathbf{x}dt &= \int_0^t \oint_S n_i q \delta v_i dS dt - \int_0^t \int_V \delta v_i (\nabla_i q) d^3\mathbf{x}dt \\ &= -\int_V [(\nabla_i q)\delta x_i]_0^t d^3\mathbf{x} + \int_0^t \int_V \left(\frac{d}{dt}\nabla_i q\right) \delta x_i d^3\mathbf{x}dt, \quad (125)\end{aligned}$$

that is

$$\int_0^t \int_V q\delta(\nabla_i v_i) d^3\mathbf{x}dt = \int_0^t \int_V (\nabla_i p) \delta x_i d^3\mathbf{x}dt, \quad (126)$$

where $p = dq/dt$ and we considered that δx_i (and δv_i as well) vanishes at the boundary S and for $t = t_1$ and $t = t_2$. Concluding, substituting (121), (124) and (126) into Eq. (119) gives:

$$\int_0^t \int_V \left(\rho \frac{dv_i}{dt} + \nabla_i p - \nabla_j P_{ji}\right) \delta x_i d^3\mathbf{x}dt = 0, \quad (127)$$

where

$$P_{ij} = -\frac{\rho}{M_w} \frac{\partial g}{\partial(\nabla_i \phi)} \nabla_j \phi \quad (128)$$

is the Korteweg stress tensor, first derived by Korteweg (1901).

Now, considering the arbitrariness of the virtual displacement δx_i and applying Reynolds theorem, we finally obtain the linear momentum equation:

$$\rho \frac{dv_i}{dt} + \nabla_i p = F_{K,i} = \nabla_j P_{ji}, \quad (129)$$

where d/dt is the material derivative, to be solved with the incompressibility constraint:

$$\nabla_i v_i = 0. \quad (130)$$

Using the expression (103) for the free energy, the Korteweg stress tensor **P** becomes:

$$P_{ij} = -\left(\frac{\rho RT}{M_w}\right) a^2 (\nabla_i \phi)(\nabla_j \phi). \quad (131)$$

Note that the Korteweg stress and force depend only on the non local part of the free energy: even when, as in Eqs. (134) and (136), this is not explicitly indicated, the thermodynamic part of the free energy contributes a term that is identically zero.

In addition, for non-dissipative systems, the heat equation and the equation of conservation of chemical species contain trivially only the convective term, that is,

$$\frac{dT}{dt} = \frac{\partial T}{\partial t} + \mathbf{v} \cdot \nabla T = 0, \tag{132}$$

$$\frac{d\phi}{dt} = \frac{\partial \phi}{\partial t} + \mathbf{v} \cdot \nabla \phi = 0, \tag{133}$$

with diffusive fluxes that are identically zero. The body force $\mathbf{F}_\phi$ in Eq. (129) can be rewritten as:

$$F_{K,i} = \nabla_j P_{ji} = -\frac{\rho}{M_w}\left[\nabla_j \left(\frac{\partial g}{\partial \nabla_j \phi}\right)\nabla_i \phi + \frac{\partial g}{\partial \nabla_j \phi}\nabla_i \nabla_j \phi + \frac{\partial g}{\partial \phi}\nabla_i \phi - \frac{\partial g}{\partial \phi}\nabla_i \phi\right],$$

that is

$$F_{K,i} = \frac{\rho RT}{M_w}\tilde{\mu}\nabla_i \phi - \frac{\rho}{M_w}\nabla_i g, \tag{134}$$

and therefore the momentum equation becomes

$$\rho\frac{dv_i}{dt} + \nabla_i p' = \frac{\rho RT}{M_w}\tilde{\mu}\nabla_i \phi, \tag{135}$$

where the pressure term has been redefined as $p' = p + (\rho/M_w)g$. Alternatively, this equation can also be written as

$$\rho\frac{dv_i}{dt} + \nabla_i p'' = -\phi\nabla_i \psi, \tag{136}$$

with a pressure term $p'' = p' - (\rho RT/M_w)\tilde{\mu}\phi$, while

$$\psi = \frac{\rho RT}{M_w}\tilde{\mu}. \tag{137}$$

Similar results were also obtained by Jasnow and Viñals (1996) and Antanovskii (1995). It should be stressed that the body force $\mathbf{F}_K$ is non dissipative, as it arises from the minimum action principle. Its expression in (136) is quite intuitive: the momentum flux is directed towards regions with smaller chemical potential differences.

The Korteweg body force can also be expressed in the following form:

$$\mathbf{F}_K = -\sum_{1=1}^{2} x_i \nabla \psi_i, \tag{138}$$

where

$$\psi_i = \frac{\rho RT}{M_w}\tilde{\mu}_{NL,i}. \tag{139}$$

is the potential energy of species i. Note that, since according to the Gibbs-Duhem relation, $\sum x_i \nabla \tilde{\mu}_{Th,i} = 0$, in the expression above $\tilde{\mu}_{NL,i}$ can be replaced by $\tilde{\mu}_i$, so that, as already seen, the Korteweg force reduces to:

$$\mathbf{F}_K = -\phi \nabla \psi, \tag{140}$$

where $\psi = \psi_1 - \psi_2 = (\rho RT/M_w)\tilde{\mu}_{NL}$. For all practical purposes, the Korteweg force can be considered as a potential force: in fact, we can include into ψ_i also the contributions of any other potential force. For example, in the Boussinesq, quasi-incompressible approximation, the buoyancy force is $\mathbf{F}_g = -\phi \nabla (\Delta\rho) g z$, where $\Delta\rho$ is the density difference between the two species, g is the gravity acceleration term and z is the vertical coordinate. Accordingly, gravity can be accounted for by simply adding the term $V_{ext} = (\Delta\rho) g z$ to ψ_i in Eq. (138). Finally, the total potential energy of the system is:

$$V = \sum_{1=1}^{2} x_i \psi_i = \frac{\rho RT}{M_w}\phi\, \tilde{\mu}_{NL} = \frac{\rho}{M_w} g_{NL} + V_{ext}, \tag{141}$$

where, as we have seen, $\tilde{\mu}_i$ includes also the contribution of all other external, potential forces.

4.2 Noether's theorem

The result (131) could be more easily determined by applying Noether's theorem. Let us illustrate this theorem for a somewhat simplified problem, where we omit the dependence of the Lagrangian $\mathscr{L}$ on $\mathbf{v}$, so that

$$\mathscr{L} = \mathscr{L}(\phi, \nabla\phi) = -\frac{\rho}{M_w} g(\phi, \nabla\phi). \tag{142}$$

Note that $\mathscr{L}$ does not depend explicitly on $\mathbf{x}$. Accordingly, the system will follow a path that is described through the Euler-Lagrange equation,

$$\frac{\partial g}{\partial \phi} - \nabla_k \frac{\partial g}{\partial(\nabla_k \phi)} = 0. \tag{143}$$

Now, consider the following equality,

$$\frac{\partial g}{\partial x_i} = \frac{\partial g}{\partial \phi}\frac{\partial \phi}{\partial x_i} + \frac{\partial g}{\partial(\nabla_k \phi)}\frac{\partial(\nabla_k \phi)}{\partial x_i}. \tag{144}$$

After multiplying the Euler-Lagrange equation by $\nabla\phi$ we obtain:

$$\frac{\partial g}{\partial x_i} - \left[\nabla_k \frac{\partial g}{\partial(\nabla_k)}\right](\nabla_i \phi) - \left[\frac{\partial g}{\partial(\nabla_k \phi)}\right](\nabla_k \nabla_i \phi) = 0, \qquad (145)$$

that is

$$\nabla_k \left\{ g\delta_{ik} - \left(\frac{\partial g}{\partial(\nabla_k \phi)}\right)(\nabla_i \phi)\right\} = 0. \qquad (146)$$

Finally, we obtain again the equation:

$$\nabla p = \nabla \cdot \mathbf{P}, \qquad (147)$$

with $p = -g\,\rho/M_w$, where $\mathbf{P}$ represents the Korteweg stresses (128) and (131). Naturally, including the kinetic term in the Lagrangian, together with the incompressibility constraint, we would obtain also the acceleration and a corrected pressure gradient term.

5 Dissipative terms

In a binary mixture, $\mathbf{v}$ denotes the mass averaged velocity between the mean velocity of component 1 and that of component 2, $\mathbf{v} = x_1\mathbf{v}_1 + x_2\mathbf{v}_2$, where we have considered that in our case mass, mole and volume average coincide.

5.1 The stress tensor

When dissipation is taken into account, the equation of motion remains basically Eq. (129), where a a momentum flux tensor $\mathbf{J}_\mathrm{v}$ (i.e. a viscous stress) is added, i.e.

$$\rho\frac{dv_i}{dt} = -\nabla \cdot \mathbf{J}_\mathrm{v} + \mathbf{F}_K, \qquad (148)$$

where, for Newtonian fluids,

$$\mathbf{J}_\mathrm{v} = p\mathbf{I} - \eta\left[\nabla\mathbf{v} + (\nabla\mathbf{v})^+\right], \qquad (149)$$

where the superscript T indicates the transpose, while η is an effective viscosity, that, assuming Newtonian behavior, is independent of the shear rate, so that $\eta = \eta(\phi)$.

5.2 The diffusive molar flux

The equation of conservation of chemical species now will contain a dis sipative term as well, that is

$$\frac{d\phi}{dt} + \nabla \cdot \mathbf{J}_\phi = r_\phi, \tag{150}$$

where $\mathbf{J}_\phi = x_1 (\mathbf{v}_1 - \mathbf{v})$ is a diffusive molar flux, with $\mathbf{v}_1$ denoting the mean velocity of species 1, while r_ϕ is the chemical reaction term, that is the number of moles of component 1 that are generated per unit volume and time,

$$r_\phi = \sum_{j=1}^{r} \nu_{1j} \mathcal{J}_j, \tag{151}$$

where ν_{kj} is the stoichiometric coefficient with which component k appears in the chemical reaction j, while $\mathcal{J}_j$ is the chemical reaction rate of reaction j. In the following, and in all relevant cases involving binary mixtures, we can assume that $r_\phi = 0$. The general case is discussed in de Groot and Mazur (1984).

As for the diffusive fluxes, when the Cahn-Hilliard part of the free energy is neglected, the mass flux of component 1 is proportional to the gradient of the chemical potential of component 1 as (Cussler, 1982, p. 180),

$$\mathbf{J}_1 = -D x_1 \nabla \mu_1, \tag{152}$$

where D is the molecular diffusivity (i.e. D is a function of temperature and pressure, but not of composition), while the proportionality term (Dx_1) has been chosen so that in the ideal case we obtain Fick's constitutive law (see below). For symmetric binary mixtures, substituting (81) into (84), we obtain the chemical potential as:

$$RT\mu_1 = g_1 + RT[\ln x_1 + \Psi x_2^2], \tag{153}$$

so that (152) yields:

$$\mathbf{J}_1 = -D x_1 \frac{d\mu_1}{dx_1} \nabla x_1 \quad \Longrightarrow \quad \mathbf{J}_1 = -D^* \nabla x_1, \tag{154}$$

where

$$D^* = D(1 - 2\Psi x_1 x_2) \tag{155}$$

is the diffusion coefficient. Inverting the suffixes 1 and 2 in (154) and (155) we see that a) the diffusivity of component 1 into 2 equals the diffusivity of component 2 into 1, as it should, and b) the flux of species 2 is opposite to the flux of species 1, that is $\mathbf{J}_2 = -\mathbf{J}_1$, showing that these are really diffusive fluxes, with no convective components. In general, from (152) and applying the Gibbs-Duhem relation, $x_1 \nabla \mu_1 = -x_2 \nabla \mu_2$, we see that it is always true that $\mathbf{J}_2 = -\mathbf{J}_1$ (Nauman and He (2001)). Therefore, for ideal or dilute mixtures, i.e. when either $\Psi = 0$ or $x_1 \ll 1$ (or $x_2 \ll 1$) we obtain that $D^* = D$ and therefore Eq. (154) reduces to Fick's law. When we plot D^* as a function of x_1 we see (see Figure 5) that for $\Psi > 2$ there is a region of negative diffusion, as it corresponds to the region where $d^2g/dx_1^2 < 0$. Going back to our notation, i.e. $x_1 = \phi$, observe that, denoting $\mathbf{J}_\phi \equiv \mathbf{J}_1$, the constitutive relation (152) can also be written as

$$\mathbf{J}_\phi = -D\phi(1-\phi)\nabla\mu, \tag{156}$$

where $\mu = \mu_1 - \mu_2$. This can be proven considering that

$$-D\phi(1-\phi)\nabla\mu_{Th} = -D\phi(1-\phi)\nabla\left(\mu_1 - \mu_2\right) = (1-\phi)\mathbf{J}_1 - \phi\mathbf{J}_2 = \mathbf{J}_1,$$

where we have considered that $\mathbf{J}_1 + \mathbf{J}_2 = 0$. Note that a Soret thermal diffusion term, $\mathbf{J}_\phi^S$, could also be added to mass flux constitutive relation, although it is generally assumed to be negligible, as discussed by Thiele et al. (2007).

At this point, a natural extension of the constitutive relation (156) is to replace the thermodynamic chemical potential with the generalized chemical potential. In fact, in Section 5.4, we see that this is indeed true and that the gradient must be taken at constant temperature, i.e. [see Eq. (178)],

$$\mathbf{J}_\phi = -D\phi(1-\phi)\left[\nabla\tilde{\mu}\right]_T, \tag{157}$$

This is the constitutive equation that has been used in Mauri et al. (1996) and in all subsequent works by Mauri and coworkers.

5.3 The energy equation

Now we impose that the time derivative of the total energy of the system equals its energy dissipation through the boundaries, i.e.

$$\frac{d}{dt}\int_V e d^3\mathbf{x} = -\oint_S \hat{\mathbf{n}}\cdot\mathbf{J}_e d^2\mathbf{x}, \tag{158}$$

where the integrals are taken over a material volume $V = V(t)$ and its delimiting surface $S(t)$. Here e is the total energy of the mixture per unit volume, i.e. the sum of kinetic, potential and internal energy,

$$e = \frac{1}{2}v^2 + V + u. \tag{159}$$

where u is the thermodynamic internal energy, which here is simply related to temperature as $u = \rho cT$, as the mixture is incompressible, while

$V = (\rho/M_w)g$ is the potential energy [see Eq. (141)] which, as we have seen, includes the non-local part of the free energy plus any other potential energy term, due to gravitational, electric, and other potential force fields. The following analysis is basically identical to that in de Groot and Mazur (1984).

First, we obtain a balance equation for the kinetic energy multiplying Eq. (148) by $\mathbf{v}$ and volume integrating:

$$\frac{d}{dt}\int_V \left(\frac{1}{2}\rho v^2\right) d^3\mathbf{x} = -\oint_S \hat{\mathbf{n}}\cdot(\mathbf{J_v}\cdot\mathbf{v})\, d^2\mathbf{x} + \int_V (\mathbf{J_v}:\nabla\mathbf{v} + \mathbf{F}_K\cdot\mathbf{v})\, d^3\mathbf{x}, \quad (160)$$

where we have integrated by parts, applying the incompressibility condition.

Then, we determine a balance equation for the potential energy, $V = (\rho/M_w)\tilde{g}_{NL}$, as:

$$\frac{d}{dt}\int_V V d^3\mathbf{x} = \int_V \left[-(\nabla\cdot\mathbf{J}_\phi)\,\psi_{NL} + \phi\mathbf{v}\cdot\nabla\psi_{NL}\right] d^3\mathbf{x}, \quad (161)$$

with $\psi_{NL} = (\rho RT/M_w)\tilde{\mu}_{NL}$. Here we have taken into account that $g_{NL} = RT(\phi\tilde{\mu}_{NL})$, $dg_{NL}/dt = RT[\tilde{\mu}_{NL}(d\phi/dt)+\phi(d\tilde{\mu}_{NL}/dt)]$, $\dot{\phi} = -\nabla\cdot\mathbf{J}_\phi$, and we have considered that $\tilde{\mu}_{NL}$ is not an explicit function of t, so that $d\tilde{\mu}_{NL}/dt = \mathbf{v}\cdot\nabla\tilde{\mu}_{NL}$. Therefore, integrating by parts we obtain:

$$\frac{d}{dt}\int_V V d^3\mathbf{x} = -\oint_S (\hat{\mathbf{n}}\cdot\mathbf{J}_\phi)\,\psi_{NL} d^2\mathbf{x} + \int_V (\mathbf{J}_\phi\cdot\nabla\psi_{NL} - \mathbf{F}_K\cdot\mathbf{v})\, d^3\mathbf{x}, \quad (162)$$

Note that the work done by the Korteweg force, as that done by any other potential force, decreases the potential energy and at the same time increases the kinetic energy by the same amount [cf. Eqs. (160) and (162)]. Accordingly, that term drops out of the balance equation for the mechanical energy:

$$\frac{d}{dt}\int_V \left(\frac{1}{2}\rho v^2 + V\right) d^3\mathbf{x} = -\oint_S \tilde{\mathbf{n}}\cdot(\mathbf{J_v}\cdot\mathbf{v} + \mathbf{J}_\phi\psi_{NL})\, d^2\mathbf{x} + \int_V (\mathbf{J_v}:\nabla\mathbf{v} + \mathbf{J}_\phi\cdot\nabla\psi_{NL})\, d^3\mathbf{x}. \quad (163)$$

Now, impose that the total energy (159) satisfies the balance equation (158), where the total energy flux, being the sum of the mechanical energy flux and the internal energy flux, can be written as

$$\mathbf{J}_e = \mathbf{J}_u + \mathbf{J_v}\cdot\mathbf{v} + \mathbf{J}_\phi\psi_{NL}, \quad (164)$$

that is the sum of, respectively, the internal energy flux (generally referred to as heat flux), the work dissipated by viscous forces, and the net potential energy loss. This last term describes the fact that when species i diffuses out of the material volume, it carries an energy $\mathbf{J}_i V_i$, where V_i is the potential energy (139) associated with component i. Finally, we derive the integral equation for the internal energy:

$$\frac{d}{dt}\int_V u\, d^3\mathbf{x} = -\oint_S \hat{\mathbf{n}}\cdot\mathbf{J}_u d^2\mathbf{x} + \int_V \left(-\mathbf{J}_\mathbf{v}{:}\nabla\mathbf{v} - \mathbf{J}_\phi\cdot\nabla\psi_{NL}\right) d^3\mathbf{x}, \qquad (165)$$

corresponding to the differential form

$$\frac{du}{dt} + \nabla\cdot\mathbf{J}_u = \dot{q}, \qquad (166)$$

where the heat generation term reads

$$\dot{q} = -\mathbf{J}_\mathbf{v}{:}\nabla\mathbf{v} - \mathbf{J}_\phi\cdot\nabla\psi_{NL}. \qquad (167)$$

Naturally, the internal energy flux, $\mathbf{J}_u$, must be expressed through a constitutive equation. Here, we will consider the simplest case, based on the consideration of next Chapter [see Eq. (177)], so that $\mathbf{J}_u$ is the sum of a Fourier heat flux and a mass diffusion term,

$$\mathbf{J}_u = -k\nabla T + \mathbf{J}_\phi h_{Th}, \qquad (168)$$

where T is the temperature of the mixture, k denotes the heat conductivity, that, in general, depends on the composition, i.e. $k = k(\phi)$, while h_{Th} is the thermodynamic partial enthalpy difference. In general, we should also add a Dufour diffusion thermal flux, which is generally negligible, as discussed by Thiele et al. (2007). In addition, the internal energy density, u, is a known function of temperature and density. Considering that density is a constant, here we will assume the simplest relation, $u = \rho c T$, where, in general, $c = c(\phi)$ is the specific heat. The same energy balance was obtained by Antanovskii (1995, 1996) by maximizing the entropy production of the system.

Integrating by parts the last term on the RHS of Eq. (165), the heat generation term can also be written as

$$\dot{q}' = -\mathbf{J}_\mathbf{v}{:}\nabla\mathbf{v} - \psi_{NL}\frac{d\phi}{dt}, \qquad (169)$$

while the internal energy flux becomes

$$\mathbf{J}'_u = -k\nabla T + \mathbf{J}_\phi\tilde{h}, \qquad (170)$$

where $\tilde{h} = h_{Th} + \psi_{NL}$ denotes the total (i.e. thermodynamic plus non-local) partial enthalpy difference, considering that the non local part of the chemical potential equals that of the enthalpy, being independent of temperature. The minus sign on the last term in the RHS of Eq. (169) is due to the fact that heat is drawn to increase the chemical energy of the mixture, i.e. when both $\tilde{\mu}$ and $\dot{\phi}$ are positive. In addition, (170) shows that the flux of the internal energy for multi-component mixtures can be written as the sum of a heat diffusive term and a heat transport term due to each diffusing species.

Note that, since $h = u+p/\rho$, the internal energy u can be replaced by the enthalpy h in the energy equation, as the dp/dt term can be neglected in any low Mach number process. As shown by de Groot and Mazur (1984), these same results are also valid for compressible, multi-component mixtures. A more recent and complete derivation can be found in Thiele et al. (2007) and Madruga and Thiele (2009).

5.4 The entropy equation

Consider the Gibbs equation for incompressible binary mixture,

$$T\frac{ds}{dt} = \frac{du}{dt} - \mu_{Th}\frac{d\phi}{dt}, \tag{171}$$

with μ_{Th} denoting the dimensional thermodynamic chemical potential difference. Substituting Eqs. (166) and (150), we obtain:

$$T\frac{ds}{dt} = -\nabla \cdot (\mathbf{J}_u - \mathbf{J}_\phi \mu_{Th}) + \dot{q} - \mathbf{J}_\phi \cdot \nabla \mu_{Th}. \tag{172}$$

Rearranging, we have:

$$\frac{ds}{dt} = -\nabla \cdot \mathbf{J}_s + \sigma, \tag{173}$$

where

$$\mathbf{J}_s = \frac{1}{T}(\mathbf{J}_u - \mathbf{J}_\phi \mu_{Th}) \tag{174}$$

is the entropy flux, while

$$\sigma = -\frac{1}{T}\mathbf{J}_s \cdot \nabla T - \frac{1}{T}\mathbf{J}_\phi \cdot \nabla \mu - \frac{1}{T}\mathbf{J}_\mathbf{v}:\nabla \mathbf{v} \tag{175}$$

is the entropy production, where $\mu = \mu_{Th} + \mu_{NL}$ is the total (i.e. thermodynamic plus non-local) dimensional chemical potential difference.

Different, albeit equivalent, expressions can also be found in de Groot and Mazur (1984), who used the equality

$$\nabla \tilde{\mu} = [\nabla \tilde{\mu}]_T - \tilde{s} \nabla T, \tag{176}$$

considering that $\tilde{h} = \tilde{\mu} + T\tilde{s}$, where $\tilde{h}$ and $\tilde{s}$ are the partial molar enthalpy difference and the partial molar entropy difference, respectively, while the subscript "T" indicates that the derivative must be taken at constant temperature. Equation (176) is valid also when $\tilde{h}$, $\tilde{\mu}$ and $\tilde{s}$ are generalized quantities, i.e. composed of both thermodynamic and non-local parts, as the non-local part of the chemical potential difference does not depend on temperature and therefore it is equal to the non-local part of the partial molar enthalpy difference, while the partial molar entropy difference is identically zero. Finally, we obtain for the entropy flux and the entropy production the following alternate expressions,

$$\mathbf{J}'_s = \frac{1}{T}\left(\mathbf{J}_u - \mathbf{J}_\phi h_{Th}\right) \tag{177}$$

and

$$\sigma' = -\frac{1}{T}\mathbf{J}'_s \cdot \nabla T - \frac{1}{T}\mathbf{J}_\phi \cdot [\nabla\tilde{\mu}]_T - \frac{1}{T}\mathbf{J}_\mathbf{v}{:}\nabla\mathbf{v}. \tag{178}$$

Considering that the entropy production term is the sum of the products between thermodynamic forces and thermodynamic fluxes, which are then related to each other through constitutive relations, this last expression of the entropy production reveals that entropy flux, $\mathbf{J}'_s$, material flux, $\mathbf{J}_\phi$, and momentum flux, $\mathbf{J}_\mathbf{v}$, must be related to the gradients of temperature, ∇T, chemical potential (thermodynamic plus non local) at constant temperature, $[\nabla\tilde{\mu}]_T$, and velocity, $\nabla\mathbf{v}$. This justifies the constitutive equations for the thermodynamic fluxes that we have used previously [see Eqs. (157) and (168)]. In fact, we have,

$$\mathbf{J}'_s = -k(\phi, T)\,\nabla T - C_1(\phi, T)\,[\nabla\tilde{\mu}]_T\,, \tag{179}$$

$$\mathbf{J}_\phi = -C_2(\phi, T)\,\nabla T - D(\phi, T)\,[\nabla\tilde{\mu}]_T\,, \tag{180}$$

$$\mathbf{J}_\mathbf{v} - p\,\mathbf{I} = -\eta(\phi, T)\left[\nabla\mathbf{v} + (\nabla\mathbf{v})^+\right], \tag{181}$$

where C_1 and C_2 are, respectively, the Dufour and the Soret, thermal diffusion, coefficients which, according to Onsager's principle, are equal to each other. Here we have assumed that the mixture is microscopically isotropic so that, according to Curie's principle, a vectorial flux, like heat or mass diffusive fluxes, cannot be coupled to a tensorial force, like shear rate. In addition, we have implicitly assumed that, according to the conservation of angular momentum, the momentum flux is a symmetric tensor, so that only the symmetric and traceless part of the shear rate stress can appear in the constitutive relation (181), where fluid incompressibility has also been accounted for.

6 Summary of the equations of motion

6.1 The equations of motion for incompressible binary mixtures

In the case of incompressible regular binary mixtures, assuming that the fluid density ρ and the molecular weight M_w are constant, the conservation equation for chemical species, momentum and internal energy are [cf. Eqs. (150), (148) and (166)]:

$$\frac{d\phi}{dt} = -\nabla \cdot \mathbf{J}_\phi, \tag{182}$$

$$\rho\frac{d\mathbf{v}}{dt} = -\nabla \cdot \mathbf{J}_\mathbf{v} + \mathbf{F}_K, \quad \nabla \cdot \mathbf{v} = 0, \tag{183}$$

$$\frac{du}{dt} = -\nabla \cdot \mathbf{J}_u + \dot{q}, \tag{184}$$

where ϕ is the molar fraction of component 1, $\mathbf{J}_\phi$ is the diffusion flux, $\mathbf{v}$ the average local velocity of the fluid mixture (in our case, molar, weight and volume averages all coincide), $\mathbf{J}_\mathbf{v}$ the viscous momentum flux, $\mathbf{F}_K$ is the Korteweg force (140), including also any other external potential force, u the internal energy density, $\mathbf{J}_u$ and $\dot{q}$ the internal energy flux and the heat generation term. In the last equation, the internal energy density u is related to temperature through a simple thermodynamic relation,

$$u = \rho c T, \tag{185}$$

where c is the specific heat. The heat generation term is [cf. Eq. (169)]

$$\dot{q} = -\mathbf{J}_\mathbf{v}{:}\nabla\mathbf{v} - \frac{\rho RT}{M_w}\tilde{\mu}\frac{d\phi}{dt}, \tag{186}$$

while the internal energy flux can be written as [cf. Eq. (170)]

$$\mathbf{J}_u = -k\nabla T + \frac{\rho RT}{M_w}\mathbf{J}_\phi \tilde{h}, \tag{187}$$

where $\tilde{h}$ is the total (thermodynamic plus non-local) partial enthalpy difference, and a Fourier constitutive relation for the heat flux has been assumed, with k denoting a heat conductivity, which is a known function of the composition ϕ. In the conservation equation for the chemical species, the material diffusive flux, $\mathbf{J}_\phi$, must be coupled to the constitutive relation [see Eq. (157)],

$$\mathbf{J}_\phi = -\phi(1-\phi)\, D\, [\nabla\tilde{\mu}]_T\,, \tag{188}$$

where D is the molecular diffusivity, the subscript "T" indicates constant temperature, and $\tilde{\mu}$ is the generalized chemical potential difference between the two species defined in (109),

$$\tilde{\mu} = \delta(\Delta g/RT)/\delta\phi. \tag{189}$$

Here Δg denotes the molar Gibbs free energy, defined as

$$\Delta g/RT = \phi \log \phi + (1-\phi)\log(1-\phi) + \Psi\phi(1-\phi) + \frac{1}{2}a^2(\nabla\phi)^2, \tag{190}$$

where R is the gas constant, a is a characteristic microscopic length and Ψ is the Margules parameter which describes the relative weight of enthalpic versus entropic forces. Since $\Psi_c = 2$ is the critical value of Ψ, we find that the single-phase region of the phase diagram corresponds to values $\Psi < 2$, while, conversely, $\Psi > 2$ in the two-phase region.

The Navier-Stokes equation (183) must be coupled with a constitutive equation for the momentum flux tensor. For a Newtonian fluid, we have [see Eq. (149)]:

$$\mathbf{J}_\mathbf{v} = p\mathbf{I} - \eta\left[\nabla\mathbf{v} + (\nabla\mathbf{v})^+\right], \tag{191}$$

with η denoting the composition-dependent fluid viscosity, while $\mathbf{I}$ is the identity dyadic and $\mathbf{A}^+$ is the transpose of $\mathbf{A}$.

The most important feature of this model is the presence in the governing equations (182)-(184) of the non-equilibrium reversible body force, $\mathbf{F}_K = \nabla\cdot\mathbf{P}$, which equals the generalized gradient of the free energy and therefore it is driven by chemical potential gradients within the mixture [see Eq. (136)],

$$\mathbf{F}_K = -\phi\nabla\psi. \tag{192}$$

where $\psi = (\rho RT/M_w)\tilde{\mu}_{NL}$, with $\tilde{\mu}_{NL} = a^2\nabla^2\phi$. In particular, when the system presents well-defined phase interfaces, such as at the late stages of phase separation, this body force reduces to the more conventional surface tension, as shown by Jasnow and Viñals (1996) and by Jacqmin (2000). Therefore, being proportional to $\nabla\tilde{\mu}$ which is identically zero at local equilibrium, $\mathbf{F}_K$ can be thought of as a non-equilibrium capillary force. Since $\mathbf{F}_K$ is driven by surface energy, it tends to minimize the energy stored at the interface resulting in a non-equilibrium attractive force between drops of the same phase, therefore driving, say, drops of the α phase towards regions of the same phase.

If other potential forces are present within the system, $\mathbf{F}_{ext} = -\phi\nabla V_{ext}$, this potential can be simply added to $\tilde{\mu}$ in Eq. (192).

Note that, when the fluid is Newtonian and with constant viscosity, Eq. (183) reduces to the simpler following equation:

$$\rho\frac{d\mathbf{v}}{dt}+\nabla p=\eta\nabla^2\mathbf{v}+\mathbf{F}_K, \qquad \nabla\cdot\mathbf{v}=0. \tag{193}$$

The ratio between convective and diffusive mass fluxes defines the Peclet number, $N_{Pe}=Va/D$, where V is a characteristic velocity, which can be estimated through (182) and (193), obtaining (Vladimirova et al., 1999a,b)

$$N_{Pe}=\frac{Va}{D}\approx\frac{a^2}{D}\frac{\rho}{\eta}\frac{RT}{M_w}\approx\frac{\sigma a}{\eta D}. \tag{194}$$

For systems with very large viscosity (e.g. polymer solutions) or very large diffusivity (e.g. mixtures very close to the critical point), N_{Pe} is small and the model describes a diffusion-driven separation process. For low-viscosity liquid mixtures, far from criticality, N_{Pe} is very large, showing that diffusion is important only in the vicinity of local equilibrium, when the body force $\mathbf{F}_K$ is negligibly small. In general, therefore, for fluid mixtures that are in conditions of non-equilibrium, either phase-separating or mixing, convection dominates diffusion. Although this approach has been developed for very idealized systems, it seems to capture the main features of real mixtures, at least during the phase separation process. This is why we did not add further terms to generalize our model, although they can be derived rather easily.

6.2 The equations of motion for one-component fluids

Very similar results can be obtained for the equations of motion of one-component systems subjected to conservative forces. As we saw in Section 2, in this case the order parameter is the density ρ, which is different in the two phases. Therefore, the mixture is not incompressible, even when each phase can be assumed to be incompressible (think, for example, to a solid-liquid phase transition). Accordingly, the mass conservation equation is the usual continuity equation for compressible fluids. As van der Waals realized, the presence of the non local term in the expression for the free energy has consequences in the determination of the remaining governing equations. As shown by Korteweg (1901), momentum conservation leads to a momentum flux tensor where, in addition to the usual viscous stresses, there appears another, so-called Korteweg, stress,

$$\mathbf{P}=\frac{RTa^2}{M_W}\mathcal{K}\left[\nabla\rho\otimes\nabla\rho-\mathbf{I}\left(\rho\nabla^2\rho+\frac{1}{2}|\nabla\rho|^2\right)\right], \tag{195}$$

where ρ is the mass density (the molar density times the molecular weight) and $\mathcal{K} = N_A d^3/M_W$ is a constant reciprocal density (of the same magnitude as the specific volume of the liquid phase). Note that only the first term is anisotropic; the rest, in effect, contributes to a modification of the fluid pressure (which is obviously irrelevant in the incompressible binary mixture case seen before). When substituted into the momentum conservation balance, this term adds the so-called Korteweg force, $\mathbf{F}_K = \nabla \cdot \mathbf{P}$, to the usual Navier-Stokes equation for fluids of non-uniform density. Physically, being proportional to the gradient of the chemical potential (with a minus sign), the Korteweg body force pushes the system towards thermodynamic equilibrium and is identically zero at equilibrium. In addition, since this force is reversible, it does not enter explicitly into the energy dissipation term. This process leads to the governing equations of a viscous compressible, non-isothermal fluid flow (see Antanovskii (1996), Lowengrub and Truskinovsky (1998), Anderson et al. (1998)):

$$\partial_t \rho + \nabla \cdot (\rho \mathbf{v}) = 0, \tag{196}$$

$$\partial_t (\rho \mathbf{v}) + \nabla \cdot (\rho \mathbf{v} \otimes \mathbf{v} + \mathbf{S} + \mathbf{P}) = \rho \mathbf{g}, \tag{197}$$

$$\partial_t (\rho e) + \nabla \cdot (\rho e \mathbf{v} + \mathbf{J}_q) + \mathbf{S} : \nabla \mathbf{v} = 0, \tag{198}$$

where e denotes the internal energy per unit mass and $\mathbf{g}$ is the gravity force/mass, $\mathbf{S}$ is the pressure-induced viscous stress tensor and $\mathbf{J}_q$ is the heat flux:[1]

$$\mathbf{S} = \left[p + \left(\frac{2}{3}\eta - \kappa \right) \nabla \cdot \mathbf{v} \right] \mathbf{I} - \eta (\nabla \mathbf{v} + \nabla \mathbf{v}^+), \tag{200}$$

$$\mathbf{J}_q = -k \nabla T, \tag{201}$$

with η and κ the fluid and bulk viscosities, respectively, and k the thermal conductivity, all known functions of density. These equations, coupled to the van der Waals equation of state and appropriate boundary and initial conditions, yield a well-posed problem.

[1] The same system of equations is arrived at when using the generalized specific free energy $\hat{e} = e + \frac{RTa^2}{2\rho M_W} \mathcal{K} |\nabla \rho|^2$, which accounts for the energy stored at the liquid-vapor interface, to enforce standard conservation principles. In that case, however, the heat flux must include an 'interstitial' contribution

$$\mathbf{J}_q = -k \nabla T - \frac{RTa^2}{M_W} \mathcal{K} \frac{d\rho}{dt} \nabla \rho, \tag{199}$$

in addition to the standard Fourier term.

7 Simulation results for incompressible binary mixtures

In this Section, we review both previously published and some new results on the isothermal mixing and demixing process as well as on heat transfer enhancement due to phase separation of incompressible regular binary mixtures. Finally, we show numerical results on isothermal spinodal decomposition of a van der Waals fluid.

Since at this stage we are interested in qualitative results, we restrict our analysis to two-dimensional systems, so that the velocity $\mathbf{v}$ can be expressed in terms of a stream function ψ, i.e. $v_x = \partial\psi/\partial y$ and $v_y = -\partial\psi/\partial x$. Consequently, the equations of motion (182) and (193) become:

$$\frac{\partial \phi}{\partial t} = \nabla\psi \times \nabla\phi - \nabla \cdot \mathbf{J}_\phi, \tag{202}$$

$$\eta\nabla^4\psi = \left(\frac{\rho RT}{M_w}\right) \nabla\tilde{\mu} \times \nabla\phi, \tag{203}$$

where

$$\mathbf{A} \times \mathbf{B} = A_x B_y - A_y B_x. \tag{204}$$

Since material transport here is diffusion-limited, the length scale of the process is the microscopic length a. Therefore, using the scaling,

$$\tilde{\mathbf{r}} = \frac{1}{a}\mathbf{r}; \qquad \tilde{t} = \frac{D}{a^2}t; \qquad \tilde{\psi} = \frac{1}{DN_{Pe}}\psi, \tag{205}$$

the equations of motion become (Vladimirova et al., 1999a,b):

$$\partial\phi/\partial\tilde{t} = N_{Pe}\tilde{\nabla}\tilde{\psi} \times \tilde{\nabla}\phi + \tilde{\nabla} \cdot \left\{\tilde{\nabla}\phi - \phi(1-\phi)\left[2\Psi + \tilde{\nabla}^2\right]\tilde{\nabla}\phi\right\}, \tag{206}$$

$$\tilde{\nabla}^4\tilde{\psi} = -\tilde{\nabla}\tilde{\nabla}^2\phi \times \tilde{\nabla}\phi, \tag{207}$$

where N_{Pe} is the Peclet number defined in (194). As noted by Lamorgese and Mauri (2002), Eq. (207) can be seen as a "static" constraint on the stream function field $\tilde{\psi}$, i.e. $\tilde{\psi} = \tilde{\psi}(\phi)$, so that the $\tilde{\psi}$-dependence on the right hand side of (206) can be formally dropped. Therefore, the Fourier-transformed system (206)-(207) can be written in the form

$$\frac{d}{dt}(e^{k^2 t}\hat{\phi}_{\mathbf{k}}) = e^{k^2 t}\hat{F}_{\mathbf{k}}, \tag{208}$$

where $\hat{\phi}_{\mathbf{k}}$ is the Fourier transform of ϕ, while the right-hand side represents the Fourier transform, $\hat{F}_{\mathbf{k}}$, of the nonlinear term of Eq. (206) multiplied by the integrating factor $e^{k^2 t}$. Note that the integrating factor allows the exact treatment of the diffusive term in Eq. (206). This system of differential equations can be time-integrated on a square domain using either the finite difference scheme described in Vladimirova et al. (2000) or the *ad hoc* pseudo-spectral method described in Lamorgese and Mauri (2002, 2005, 2006). In this latter case, the nonlinear term on the right hand side of Eq. (207) would normally require five FFTs for its pseudospectral evaluation. However, using the identity

$$\nabla\phi \times \nabla\nabla^2\phi = \partial^2_{xy}(\phi_x^2 - \phi_y^2) + (\partial_y^2 - \partial_x^2)\phi_x\phi_y, \tag{209}$$

its computation requires only four FFTs. As a result, it is easy to see that each time step (e.g. assuming a simple Eulerian scheme) in (207) requires the evaluation of at least thirteen FFTs.

7.1 Mixing of regular mixtures

In this Section we present the simulation results of the mixing process that a viscous and macroscopically quiescent binary mixture undergoes when it is instantaneously brought from the two- to the one-phase region of its phase diagram. In addition to presenting some new data, we will also summarize the main results by Vladimirova and Mauri (2004) and Lamorgese and Mauri (2006).

First, let us describe the mixing process between two fluids which are initially quiescent and separated by a plane interface, $r_1 = L/2$. In this case, the RHS of Eq. (207), i.e. the Korteweg body force, is identically zero, so that $\mathbf{v} = 0$ and therefore the process does not depend on the Peclet number. In fact, Eq. (206) (in its dimensional form) is well approximated by the following equation:

$$\frac{\partial\phi}{\partial t} = D^* \frac{\partial^2\phi}{\partial r_1^2}, \tag{210}$$

with

$$D^* = D[1 - 2\Psi\bar{\phi}(1-\bar{\phi})], \tag{211}$$

where $\bar{\phi}$ represents the mean value of ϕ, as the neglected terms play a role only at the very beginning of the mixing process, when the interface is still sharp. As shown in Figure 5, the results of our simulations are in perfect agreement with the similarity solution resulting from Eq. (210). This confirms that the mixing process of two fluids separated by an initially plane sharp interface remains one-dimensional, does not depend on the Peclet number and is a purely diffusive process, with an effective diffusivity D^*

that depends on the thermodynamic properties of the mixture, i.e. the value of the Margules parameter. The same result is obtained whenever the initial configuration is one-dimensional, as in the case of an isolated drop.

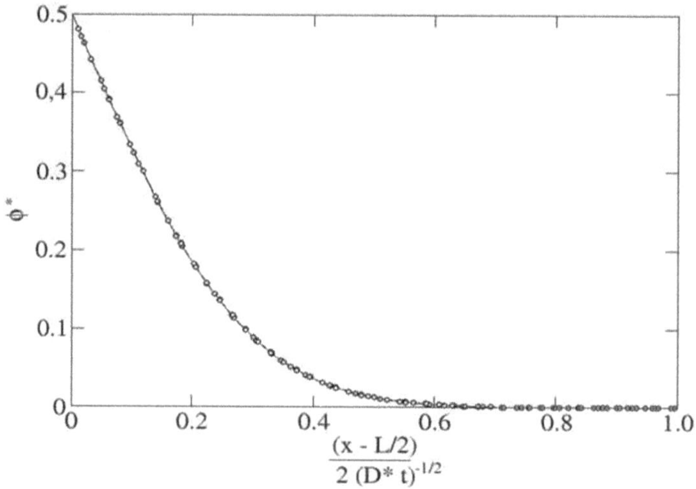

Figure 5. Simulation result of the mean composition of the mixture as a function of position and time. The continuous line represents the conjugate error function, i.e. the exact similarity solution of Eq. (210).

When we simulate the mixing process of a collection of drops immersed in a background field, it has been shown (Vladimirova et al., 1999a) that, while a single drop remains still as it is absorbed, two drops tend to attract each other and even coalesce, provided that the Peclet number is large enough and the drops are initially very close to each other. This effect is further investigated in the simulations shown in Figures 6, representing the evolution of eight identical spherical drops with radius $2a$, that are placed within a quiescent bulk fluid with $\Psi = 1.9$, at a $2a$ distance from each other. When $N_{Pe} = 0$, the drops do not move while they are reabsorbed by diffusion; on the other hand, when $N_{Pe} = 10^3$ they rapidly coalesce and form a larger isolated single drop. This larger drop, though, has to be eventually reabsorbed by diffusion and, being larger than the original drops,

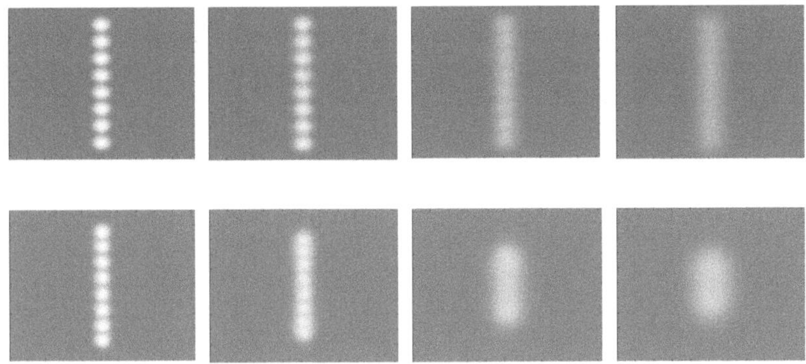

Figure 6. Evolution of eight identical drops with $\Psi = 1.9$ placed on a straight line at nondimensional times $t = 0.01, 0.02, 0.04, 0.1$ with $N_{Pe} = 0$ (top) vs. $N_{Pe} = 10^3$ (bottom).

will take approximately eight times as long to disappear. Consequently, mixing appears to be faster in the absence of convection. This result is confirmed by simulating the mixing process of a random distribution of 250 drops with radii between $5a$ and $20a$, immersed in a quiescent continuous bulk fluid with $\Psi = 1.9$. The results of these simulations are represented in Figure 7 in terms of the degree of mixing δ_m, defined as

$$\delta_m(t) = \frac{\langle|\phi(\mathbf{r},t) - \phi_{av}|^2\rangle}{\langle|\phi_0(\mathbf{r}) - \phi_{av}|^2\rangle}. \tag{212}$$

We see that δ_m decays exponentially as

$$\delta_m = \exp\left(-10^3 \frac{Dt}{a^2\tau}\right), \tag{213}$$

where τ is a non-dimensional decay time, with $\tau = 0.8$ when $N_{Pe} = 0$ and $\tau = 2.1$ when $N_{Pe} = 10^4$. This result shows that the degree of mixing decays faster in the absence of the convection that is induced by non-equilibrium interfacial forces, i.e. for very viscous liquids. The explanation of this result is that in low-viscosity liquid systems the Korteweg stresses induce a material flux which is much larger than that due to pure molecular diffusion. Therefore, as coalescence among drops is strongly enhanced, drops tend to form larger domains, which, however, eventually must dissolve by diffusion.

Therefore, in the absence of any external agitation, convection effectively slows down mixing, which therefore, contrary to common thinking, is faster when the mixtures have large viscosities.

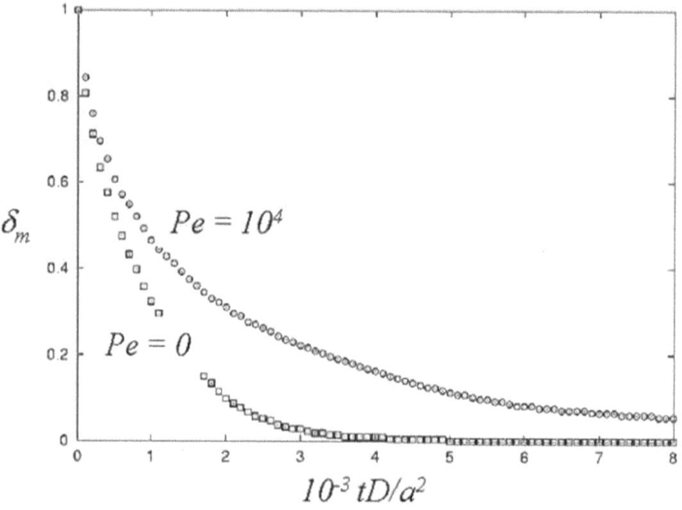

Figure 7. Degree of mixing (183) as a function of time when the Peclet number is 0 and 10^4.

The movement of isolated drops during phase transition has been observed experimentally by Molin et al. (2007) and Poesio et al. (2009), in good agreement with the predictions by Vladimirova et al. (1999a).

7.2 Spinodal decomposition of regular mixtures

In this Section we present the simulation results of Lamorgese and Mauri (2002, 2008) about the spinodal decomposition process that a viscous and macroscopically quiescent binary mixture undergoes when it is instantaneously brought from the one- to the two-phase region of its phase diagram, deep within the spinodal region. In Figure 8 we show some typical patterns of the spinodal decomposition process of slightly off-critical mixtures having uniform initial composition $\phi_0 = 0.45$ and for different values of the Peclet number, N_{Pe}, when they are instantaneous quenched well below their

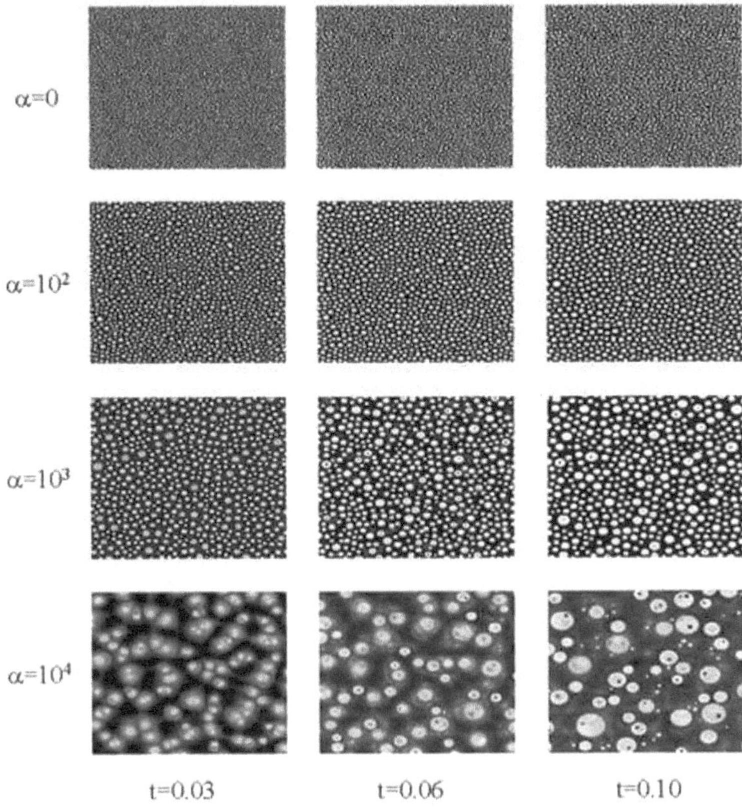

Figure 8. Phase separation patterns of four off-critical binary mixtures with mean composition $\phi_0 = 0.45$, Margules parameter $\Psi = 2.1$ and Peclet numbers $N_{Pe} = 0, 10^2, 10^3$ and 10^4 at different non-dimensional times Dt/a^2.

critical temperature, i.e. with $\Psi = 2.1$. We chose this particular value of Ψ because that is the Margules parameter of the water-acetonitrile-toluene mixture at $20°C$ that we used in the our experimental work (Gupta et al., 1996, 1999; Santonicola et al., 2001; Califano and Mauri, 2004; Poesio et al., 2006). The first row of images represents the results for $N_{Pe} = 0$, i.e. for the case when diffusion is the only mechanism of mass transfer, showing that, soon after the first drops appear, they coalesce into drop-like structures (in the critical case, we observe dendritic structures, instead). The mean composition within (and without) these structures changes rapidly, as at time $t = 0.03$ we already see two clearly distinguishable phases with almost uniform concentrations equal to 0.59 and 0.41, while at equilibrium their respective compositions are $\phi_{eq}^{\alpha} = 0.685$ and $\phi_{eq}^{\beta} = 0.315$. After this early stage, the structures start to grow, increasing their thickness and reducing the total interface area, while at the same time the composition within the domains approaches its equilibrium value. This, however, is a slow process, driven only by diffusion, and at time $t = 0.1$ the phase domains still have a drop-like geometry with a characteristic size which is just about twice as large as its value at $t = 0.01$. In the following, we will denote these slow-changing configurations as metastable states, referring to Vladimirova et al. (1998) for further information on their evolution. For non-zero convection, i.e. for $N_{Pe} \cong 0$, dendritic structures thicken faster, but up to $N_{Pe} \cong 10^3$ domain growth still follows the same pattern as for $N_{Pe} = 0$: first, single-phase domains start to appear, separated from each other by sharp interfaces, and only later these structures start to grow, with increasing growth rate for larger N_{Pe}. When $N_{Pe} > 10^3$, however, phase separation occurs simultaneously with the growth process. For example, when $N_{Pe} = 10^4$, we see the formation of isolated drops of both phases, surrounded by the bulk of the fluid mixture, which is still not separated. In addition, drops appear to move fast and randomly while they grow, absorbing material from the bulk, colliding with each other and coalescing, so that single-phase domains grow much faster than when molecular diffusion is the only transport mechanism. Considering, for example, the morphology of the system at time $t = 0.1$ for $N_{Pe} = 10^4$ and $N_{Pe} = 0$, we see that, while in the first case single-phase domains have already reached a size comparable to that of the container's, in the absence of convection the dendritic domains have an approximate widths of $20a$. Clearly, since the motion of the interface is too quick for the diffusion of concentration to establish a metastable state within the microdomains, double, or multiple, phase separation is observed, in agreement with previous numerical (Jasnow and Viñals, 1996; Tanaka and Araki, 1998; Vladimirova et al., 1999b) and experimental (Tanaka, 1996; Gupta et al., 1999) results. As a quantitative characterization of the influence of the convection parameter N_{Pe} on the average phase composition within the phase domains, we defined the separation depth, s, measuring the "distance" of the single-phase domains from their equilibrium state, i.e.,

$$s = \left\langle \frac{\phi(\mathbf{r}) - \phi_0}{\phi_{eq}(\mathbf{r}) - \phi_0} \right\rangle, \tag{214}$$

where ϕ_0 is the initial composition, and the bracket indicates volume and ensemble average. Here, ϕ_{eq} is the steady state composition of the α-phase, $\phi_{eq}^{\alpha} = 0.685$ or the β-phase, $\phi_{eq}^{\beta} = 0.315$, depending on the local composition ϕ, i.e.,

$$\phi_{eq}(\mathbf{r}) = \begin{cases} \phi_{eq}^{\alpha} & \text{when } \phi(\mathbf{r}) > \phi_0, \\ \phi_{eq}^{\beta} & \text{when } \phi(\mathbf{r}) < \phi_0. \end{cases} \tag{215}$$

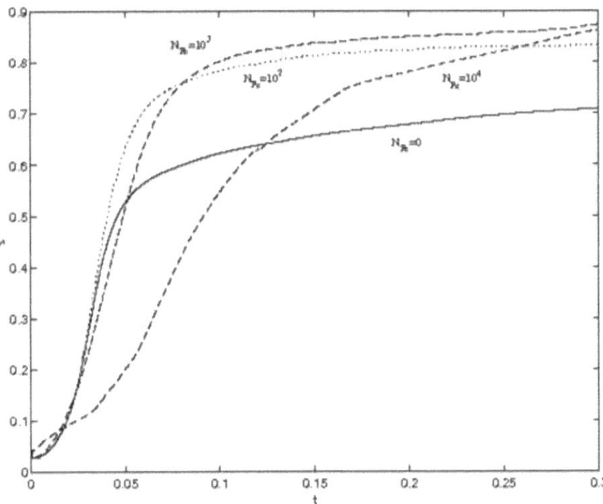

Figure 9. Separation depth (188) as a function of the non-dimensional time for Peclet numbers $N_{Pe} = 0, 10^2, 10^3$ and 10^4.

In Figure 9 the separation depth s is plotted as a function of time, using $N = 512$ modes. As we see, after an off-critical quench no detectable phase separation takes place until $t \cong 0.02$, when the first spinodal decomposition pattern is formed. Then, phase separation can take place in two different ways, depending on whether $N_{Pe} < 10^3$ or $N_{Pe} > 10^3$. For smaller N_{Pe}, single-phase domains develop very rapidly, until, at $t \cong 0.06$, they appear to be separated by sharp interfaces. From that point on, separation proceeds much more slowly, as the concentration gradients within the single-phase domains are very small, while the concentrations of the two phases across any interface change only slowly in time, asymptotically approaching quasi-equilibrium with $s = 1$. On the other hand, for $N_{Pe} > 10^3$, the growth of the separation depth is more gradual, revealing that separation and growth occur simultaneously, although obviously, at some later stage, sharp interfaces will eventually appear even in this case. Therefore, it appears that local equilibrium (with $s = 1$) is very rarely achieved for low-viscosity liquid mixtures. In this case, in fact, $a \approx 10^{-5} cm$ and, consequently, our domain size corresponds to $100\,\mu m$. Therefore, since for most liquid mixtures drops start sedimenting when they reach $100\,\mu m$ size, the system becomes gravity driven and rapidly separates before reaching the scaling regime (i.e. when $s = 1$). As shown in Lamorgese and Mauri (2002), the typical drop size grows linearly in time during the first stage of the process, i.e. until sharp interfaces are formed, while during the last stage it grows like $t^{1/3}$. Note that the behavior of a phase-separating system depends as much on the driving force $\mathbf{F}_K$ as on the Peclet number, N_{Pe}, as $\mathbf{F}_K$ (which is a function of the separation depth) can induce a strong convection only for systems with small viscosities (i.e. large N_{Pe}'s), while for very viscous systems it has hardly any effect. Very similar results were obtained by Vladimirova et al. (1999b), who used the same model but a very different numerical scheme (i.e. finite difference instead of pseudo-spectral). As shown by Lamorgese and Mauri (2008), comparison between 2D and 3D results reveals that 2D simulations capture, even quantitatively, the main features of the phenomenon.

7.3 Homogeneous nucleation of regular mixtures

In thermodynamics (Sandler, 1999), an equilibrium state (i.e. one that does not change in time while the system is isolated) is stable when it can be altered to a different state only by perturbations that do leave net effects in the environment of the system, e.g. when its temperature is changed. On the other hand, states of unstable and metastable equilibrium are equilibrium states that may be changed to different states by means of perturbations that leave no net effects on the environment. Depending on whether such perturbations are infinitesimal or finite, the system is in a state of unstable or metastable equilibrium, respectively. As we saw in Section 3, in the case of a partially miscible binary mixture the boundary of the region of stable equilibrium defines the miscibility curve in the temperature - mole fraction phase diagram $T - \phi$ at constant pressure. In addition, as an unstable system will transform spontaneously to its more stable state, Gibbs (1876) showed that in those conditions the molar Gibbs free energy g of the binary mixture must be a concave function of ϕ, i.e. $\partial^2 g(T,P)/\partial\phi^2 < 0$. In particular, the boundary of the unstable region is defined by the locus $\partial^2 g(T,P)/\partial\phi^2 = 0$, which is called the spinodal curve. Therefore, the points on the phase diagram comprised between the miscibility and the spinodal curves define the metastable region. In his classical treatment on stability, Gibbs distinguished two types of perturbations that can be applied to a homogeneous system: the first is small in intensity but large in extent, as exemplified by a small composition fluctuation spread over a large volume, while the second is large in intensity but small in extent, as it happens in a nucleation process. An unstable system is best studied assuming that it is perturbed through a delocalized, infinitesimal fluctuation in composition. In this case, in fact, the problem can be linearized, showing that the intensity of any mode whose wavelength is larger than a critical value grows exponentially in time, with the maximum growth corresponding to the typical length scale of the phase separation process. Now, at this point it would seem logical to study metastable systems by perturbing them with delocalized, although finite, composition fluctuations, so that their behavior could be easily compared with that of unstable systems. Instead, Gibbs chose to use the other type of perturbation, which is large in intensity but small in extent, assuming that a uniform droplet of the minority phase would appear within the majority phase. In this way, using the concept of surface tension between two phases at thermodynamic equilibrium that he had developed, Gibbs was able to show that while, predictably, any nucleus, even an infinitesimal one, would grow spontaneously when the mixture is unstable, metastable systems become unstable only when the nucleus exceeds a critical size. Now, apart from the fact that, as Gibbs himself recognized, it is not very reasonable to assume that a small nucleus could be homogeneous, there is not a good reason today why we should study unstable and metastable systems using different procedures. In fact, Lamorgese and Mauri (2005) studied non-stable binary mixtures, both unstable and metastable, by perturbing them with delocalized random fluctuations. Although this idea is not new (see the review article by Gunton, 1999, and references therein), they were the first to show that, as the mixture composition approaches its value at the coexistence curve, the intensity of the perturbation that is needed to trigger the instability grows exponentially, as shown in Figure

10. In fact, using a pseudo-spectral method, they simulated the nucleation process, showing that the metastability of the system can be characterized through either a critical radius, as in Gibbs' treatment, or the (finite) intensity of a white noise superposed on the initial uniform concentration field; this critical intensity grows exponentially as the mean composition of the mixture approaches its equilibrium value (see Figure 10). In addition, they showed that, in general, the value of the critical radius decreases as the number density of the nucleating drops becomes very large, so that nuclei have the chance to coalesce and grow before being reabsorbed.

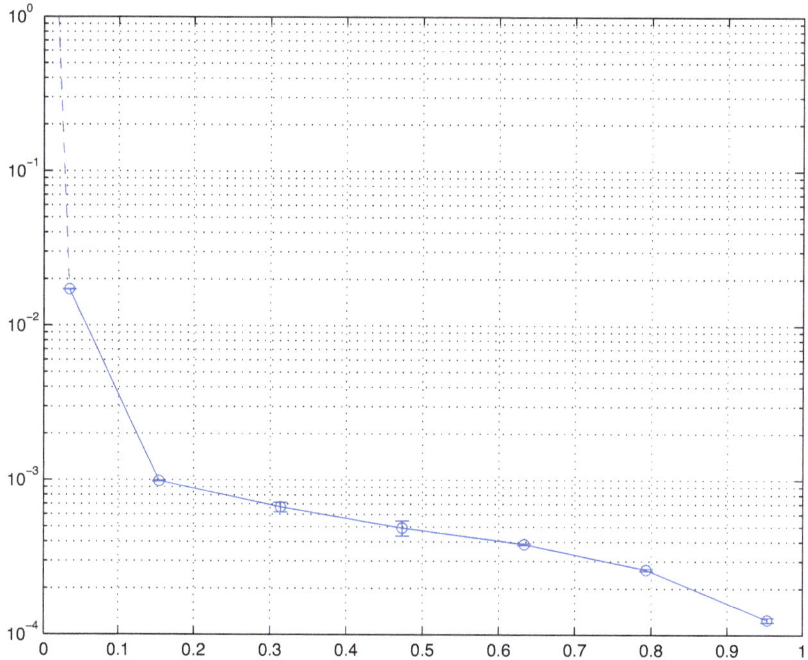

Figure 10. Critical value of the white noise perturbation, $\widetilde{\phi}^*$, as a function of the metastability degree $\xi = [\phi_0 - \phi_s]/[\phi_{eq} - \phi_s]$.

7.4 Effect of phase separation on heat transfer

The effect of temperature on phase separation has not been studied very carefully so far, with the exception of spinodal decomposition of very viscous mixtures, where convection can be neglected and, therefore, the temperature field is uncoupled to the velocity and concentration fields. In that case, Vladimirova et al. (1998) showed that, as expected, the mixture starts to separate within the region where the temperature has crossed the miscibility curve. In addition, critical mixtures show a wall effect, i.e. the dendrites tend to align along the temperature gradient, while in the absence of temperature gradients they have random directions. The most interesting case, though, is that of low-viscosity, regular mixtures, where the process is convection-driven. In that case, considering that $\Psi = 2T_c/T$ and $a = \hat{a}\sqrt{\Psi}$, where $\hat{a}$ is a constant independent of T, Eq. (182), (184) and (188) become:

$$\frac{\partial \phi}{\partial t} = N_{Pe}\tilde{\nabla}\tilde{\Psi} \times \tilde{\nabla} \cdot (\tilde{\nabla}\phi - \phi(1-\phi)[\Psi(2+\tilde{\nabla}^2)\tilde{\nabla}\phi + (2\phi - 1)\tilde{\nabla}\Psi]), \quad (216)$$

$$\tilde{\nabla}^4\tilde{\psi} = -\tilde{\nabla}(\Psi\tilde{\nabla}^2\phi) \times \tilde{\nabla}\phi, \quad (217)$$

$$\frac{\partial \Psi}{\partial t} = -N_{Pe}\tilde{\nabla}\tilde{\psi} \times \tilde{\nabla}\Psi + N_{Le}[\tilde{\nabla}^2\Psi - \frac{2}{\Psi}(\tilde{\nabla\Psi})^2], \quad (218)$$

where we have used the scaling (205), with $\hat{a}$ replacing a and N_{Le} denotes the Lewis number, $N_{Le} = \alpha/D$. In Eq. (218) we have neglected the heat transport induced by mass diffusion and the heat generation term of the general energy conservation equation, assuming that the adimensional specific heat is negligibly large, i.e. $\hat{c} = c/R \gg 1$. As we have repeatedly stressed, in Eq. (218) the temperature field is coupled to the velocity field and therefore the temperature distribution cannot be imposed. This case was studied by Molin and Mauri (2007), showing how the temperature field is coupled to the velocity field. In all cases, the mixture starts to phase separate at the walls; then, as heat losses penetrate deeper within the domain, demixing takes place everywhere, until, at steady state, the temperature of the mixture reaches its equilibrium value. As shown in Fig. 11, a) the mixture phase separates through the formation of bicontinuous structures; b) in the presence of convection, i.e. when $N_{Pe} = 10^2$, warm fluid tends to move towards the wall, thus enhancing heat transport.

The most obvious way to describe the heat transfer enhancement is through the Nusselt number, which is defined as the ratio between the heat flux J_q at the wall and the heat flux that one would have in the absence of convection, $(J_q)_{N_{Pe}=0}$, i.e.,

$$N_{Nu} = J_q / (J_q)_{N_{Pe}=0} \, . \quad (219)$$

Although N_{Nu} would seem to be a function of time, we saw that it actually remains almost constant and therefore can be used effectively to characterize the enhancement of heat transport due to phase transition. The most important result of our simulation was to determine how such

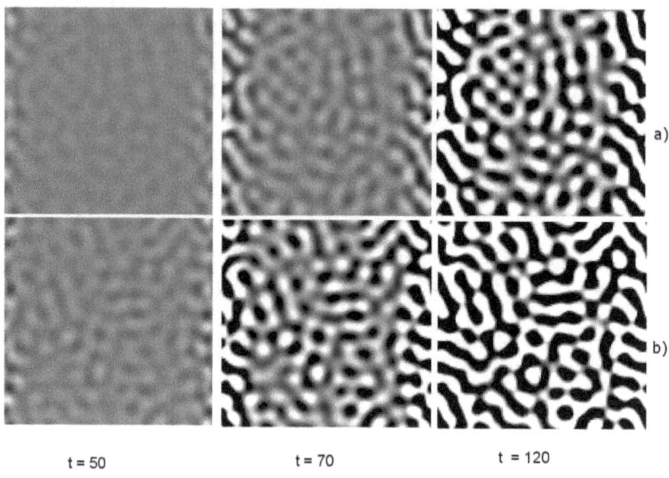

Figure 11. Evolution of the concentration field for $N_{Pe} = 0$ (top) and $N_{Pe} = 10^2$ (bottom) when $N_{Le} = 1$, $\zeta = 0.05$ and with no source term. The snapshots are taken at $t = 50$, $t = 70$ and $t = 120$, where t is expressed in $\hat{a}^2/D$ units.

heat transport enhancement depends on the characteristics of the process, namely the Peclet and Lewis numbers, the specific heat and the quenching depth. In particular, in Figure 12 we see that heat transport increases monotonically with N_{Pe} until it reaches a plateau at $N_{Nu} \approx 2.0$ when $N_{Pe} \approx 10^6$. This result is in qualitative agreement with the experimental data by Poesio et al. (2007).

8 Spinodal decomposition of a van der Waals fluid

We discuss below simulation results of Lamorgese and Mauri (2009) on isothermal liquid-vapor spinodal decomposition of a van der Waals fluid, occurring after quenching it from a single-phase equilibrium state to the unsteady two-phase region of its phase diagram, showing that the process is governed by the density gradient-driven non-equilibrium Korteweg body

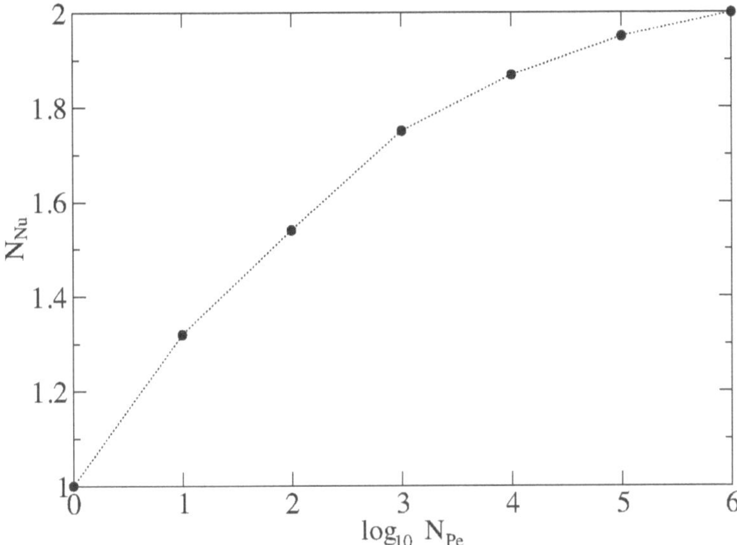

Figure 12. The Nusselt number, N_{Nu}, as a function of the Peclet number, N_{Pe}, when $N_{Le} = 1$, $\zeta = 0.05$ and with no source term.

force. Previously, liquid-vapor phase separation has been investigated by experiment and simulation (Yamamoto and Nakanishi, 1996; Beysens et al., 2002; Borcia and Bestehorn, 2007; Oprisan et al., 2008), but none of these previous works has focused on spinodal decomposition systematically. Following the diffuse-interface model (Felderhof, 1970; Antanovskii, 1996; Anderson et al., 1998), Lamorgese and Mauri (2009) investigate liquid-vapor spinodal decomposition in 2D and 3D for critical and off-critical van der Waals fluids as a function of a convection parameter expressing the relative magnitude of capillary to viscous forces, showing that, at the late stages of the process, the mechanism of growth is convection-driven coalescence with a $\frac{2}{3}$ power-law scaling for the characteristic size of single-phase microdomains, in agreement with dimensional analysis (Siggia, 1979; Furukawa, 1994) and experimental measurements (Beysens et al., 2002).

The case of an off-critical van der Waals fluid is presented below as more complete results have been reported elsewhere (Lamorgese and Mauri, 2009). There, the equations of motion were obtained rigorously coupling the van der Waals equation of state with the fundamental conservation principles and invoking irreversible thermodynamics (cf. Section 6.2). Also, the governing equations were made dimensionless based on both a diffusive scaling (which is relevant to the beginning of phase separation) and an acoustic, or convective, scaling (which is relevant to the late stage of phase separation). We recall basic definitions of non-dimensional groups that emerge under the acoustic scaling. Accordingly, defining a reference speed of sound, $u_s = (RT_C/M_W)^{1/2}$, we set:

$$\tilde{\mathbf{x}} = \frac{\mathbf{x}}{L}, \quad \tilde{t} = \frac{u_s t}{L}, \quad \tilde{\mathbf{u}} = \frac{\mathbf{u}}{u_s}, \quad Re = \frac{\rho_c u_s L}{\eta_I}, \quad \beta = \frac{N_A d^3}{M_W}\rho_C, \tag{220}$$

$$\tilde{\rho} = \frac{\rho}{\rho_C}, \quad \tilde{p} = \frac{p}{\rho_C RT_C/M_W}, \quad \tilde{T} = \frac{T}{T_C}. \tag{221}$$

Here, d, N_A and M_W denote a molecular diameter, the Avogadro number and the molecular weight, respectively, whereas ρ_C and T_C are the critical density and temperature, and $p_C^* = \rho_C RT_C/M_W$ is the pressure at critical conditions if the fluid were a perfect gas (R being the gas constant), and β is an $O(1)$ density ratio. Re is an acoustic Reynolds number which can be used to define an interfacial Reynolds number as

$$\mathscr{R} = (ReCn)^2 = \left(\frac{\rho_c u_s a}{\eta_I}\right)^2, \tag{222}$$

where η_I is the viscosity of the liquid phase and where the Cahn number, $Cn = a/L$, is the ratio of the interface thickness a to the macro length scale L. In fact, it can be shown that the non-dimensional parameter $\mathscr{R}$ can be defined independently as the ratio of capillary-to-viscous forces (i.e. as an inverse capillary number). In the governing equations, all external forces, and buoyancy in particular, were assumed to be negligible. In other words, drop and bubble sizes ℓ are small compared to the capillary length, i.e., $\ell \ll \sqrt{\sigma/g\,(\rho_I - \rho_{II})}$ (with σ the surface tension, g the gravity field and with I and II denoting the liquid and vapor phases at equilibrium), which amounts to assuming that the Bond number is very small. Also, for the sake of simplicity, we assumed $\beta = 1$.

Numerical methods in Lamorgese and Mauri (2009) are based on previous works by Nagarajan et al. (2003, 2007). Accordingly, the governing equations were discretized using a structured, staggered arrangement of the conserved variables, with spatial derivatives computed via sixth-order compact finite differences (Lele, 1992). Temporal advancement was effected via a third-order Runge-Kutta scheme.

We investigated liquid-vapor spinodal decomposition of a van der Waals fluid which is instantaneously quenched into the unstable region of its phase diagram. The simulations were conducted at $\tilde{T} = 0.9$, for negligible Bond number and for different values of $\mathscr{R}$, assuming an infinite expanse of fluid (modeled via periodic boundary conditions). At this temperature, a representative vapor-liquid viscosity ratio was chosen to be $r = 10^{-3}$. Initially, we assumed quiescent conditions with a density field being the sum of a Gaussian white noise superposed on a uniform constant density $\rho = \rho_0$. We chose values of $\tilde{\rho}_0 = 0.7$, 1.042, 1.3 in the spinodal range, i.e., $\tilde{\rho}_0 \in [\tilde{\rho}_s^{II}, \tilde{\rho}_s^{I}]$, where $\tilde{\rho}_s^{I} = 1.39$ and $\tilde{\rho}_s^{II} = 0.654$ denote the liquid and vapor spinodal densities at the given temperature. (For a van der Waals fluid these values are easily found by solving the algebraic relation $\tilde{\rho}(3 - \tilde{\rho})^2 = 4\tilde{T}$.) Specifically, $\tilde{\rho}_0 = 1.042$ is the critical density, while $\tilde{\rho}_0 = 0.7$ and $\tilde{\rho}_0 = 1.3$ correspond to vapor-rich and liquid-rich mixtures, respectively. As expected, we observed that the phase-ordering process after the critical quench $\tilde{\rho}_0 = 1.042$ is characterized by the formation of bicontinuous structures, which subsequently grow and coalesce. For the off-critical quench $\tilde{\rho}_0 = 1.3$ (Fig. 13), the phase separation pattern consists of a random collection of rapidly coalescing pseudo-spherical nuclei of the minority phase, surrounded by the majority phase. Only after the first spinodal pattern is formed (i.e. a bicontinuous pattern for the critical quench, or a random collection of nuclei for the off-critical quench), do the single-phase domains start to grow and coalesce, at an increasing rate for larger $\mathscr{R}$.

As a quantitative characterization of the influence of the convection parameter $\mathscr{R}$ on the average phase composition of the two-phase fluid, we defined the separation depth, s, measuring the "distance" of the single-phase domains from their equilibrium state (Vladimirova et al., 1998) as

$$s = \left\langle \frac{\rho(\boldsymbol{x}) - \rho_0}{\rho_{eq}(\boldsymbol{x}) - \rho_0} \right\rangle, \tag{223}$$

where ρ_0 is the initial mean density, and the brackets indicate volume and ensemble averaging. Here, ρ_{eq} denotes the steady-state density of the liquid phase, ρ_I, or of the vapor phase, ρ_{II}, depending on the local density $\rho(\boldsymbol{x})$, i.e.,

$$\rho_{eq}(\boldsymbol{x}) = \begin{cases} \rho_I, & \text{if } \rho(\boldsymbol{x}) > \rho_0, \\ \rho_{II}, & \text{if } \rho(\boldsymbol{x}) < \rho_0. \end{cases} \tag{224}$$

Figure 14 shows the temporal evolution (in acoustic time units) of the separation depth for the off-critical quench $\tilde{\rho}_0 = 1.3$ for $\mathscr{R} = 1$, 10, 100, 1000. The solid curves in this figure were obtained from 2D simulations on a 256^2 grid, while the dotted lines are from 3D simulations on a 128^3 grid. This figure shows remarkable quantitative agreement between the two sets of curves. Next, we studied the rate of coarsening as reflected in the growth

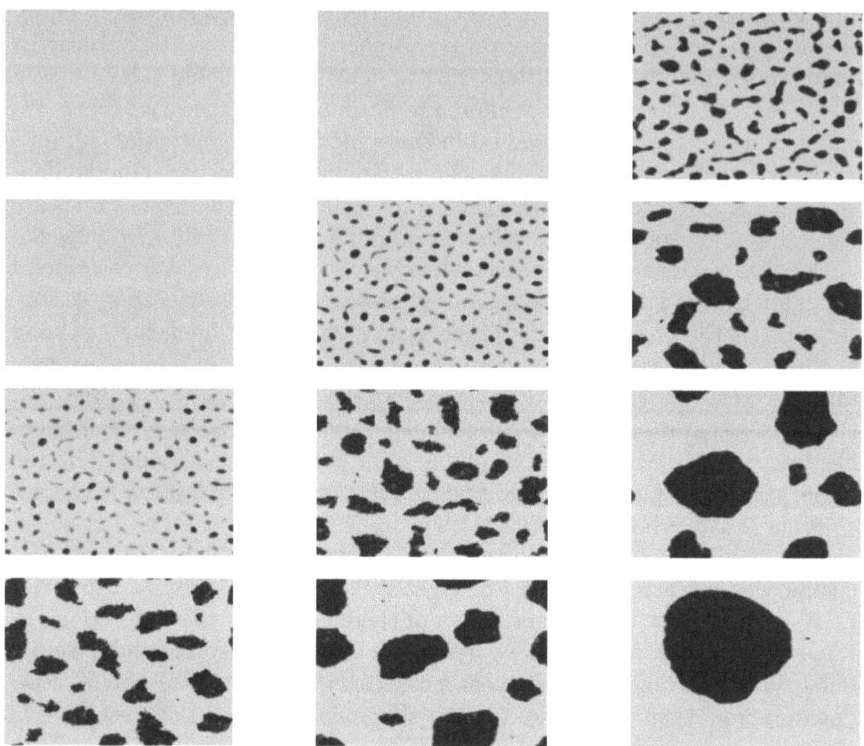

Figure 13. Liquid-vapor spinodal decomposition of an off-critical van der Waals fluid (with $\tilde{\rho}_0 = 1.3$) at different non-dimensional times (diffusive scaling) $\tilde{t} = 3\,10^{-3}$, 10^{-2} and $5\,10^{-2}$, with $\mathscr{R} = 1$, 10, 10^2 and 10^3 from top to bottom.

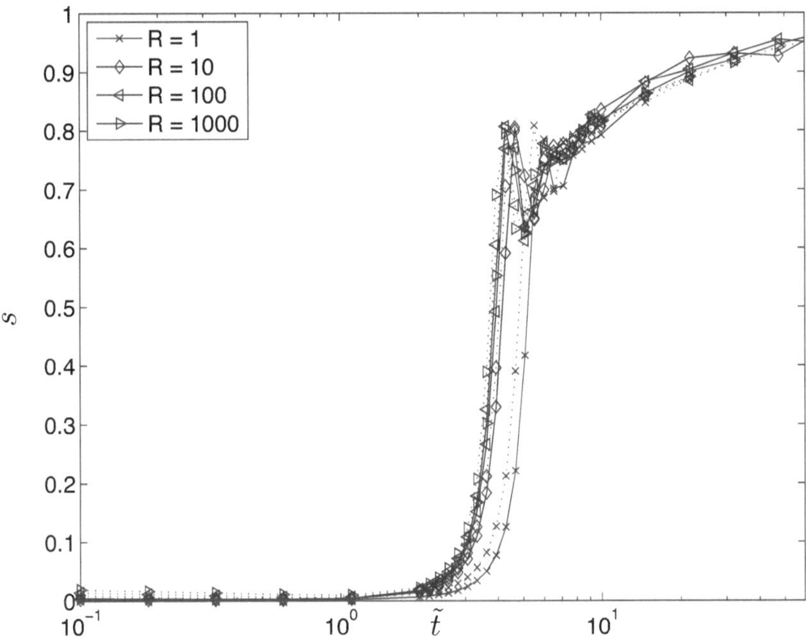

Figure 14. Separation depth vs. time (acoustic scaling) for different values of $\mathscr{R} = 1, 10, 100, 1000$ from 2D simulations with $\tilde{\rho}_0 = 1.3$ (solid) vs. 3D simulation results (dotted).

law for the integral scale

$$\mathcal{L}(t) = \frac{1}{\rho_{rms}^2} \sum_{\mathbf{k}} \frac{\langle |\hat{\tilde{\rho}}_{\mathbf{k}}|^2 \rangle}{|\mathbf{k}|}, \tag{225}$$

where $\tilde{\rho} = \rho - \langle \rho \rangle$, ρ_{rms} is the root-mean-squared value of ρ, hats denote Fourier transforms, while the brackets denote averaging over a shell in Fourier space at fixed $k = |\mathbf{k}|$. As can be seen in Fig. 15, during the first stage of the process the rate of coarsening is strongly influenced by the chosen value for the convection parameter, until sharp interfaces are formed. Then, domains stop growing, concomitant to their composition rapidly reaching local equilibrium. Finally, during the latest stage, growth is driven by inertial forces and is characterized by a $\frac{2}{3}$ power-law behavior, in agreement with predictions based on simple dimensional analysis (Siggia, 1979; Furukawa, 1994). So, first there is a time delay, with no de-

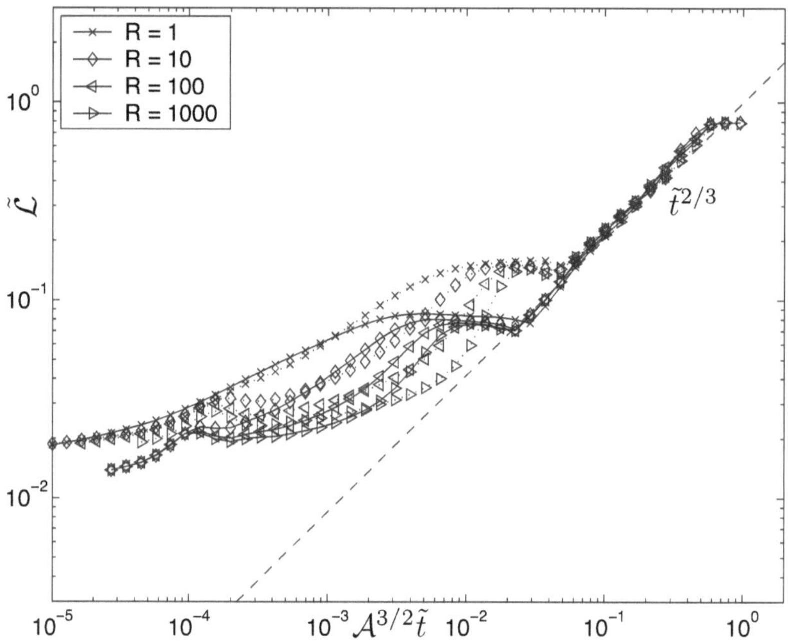

Figure 15. Integral scale vs. time (acoustic scaling) for different values of $\mathscr{R} = 1$, 10, 100, 1000 from 2D simulations with $\tilde{\rho}_0 = 1.3$ (solid) vs. 3D simulation results (dotted). $\mathcal{A}$ is the non-dimensional $(\sigma/\rho)^{1/3}$ ratio, based on the scaling theory growth law $\tilde{\mathcal{L}} = \mathcal{A}\tilde{t}^{2/3}$.

tectable liquid-vapor phase separation. Then, during the second and third stages, first the system reaches local equilibrium (with the formation of nuclei having sharp interfaces), and then these nuclei start to grow. This is in contrast with phase separation in viscous liquid binary mixtures, where the two events occur simultaneously (cf. discussion in Section 7.2). There, after the (delayed) onset of phase separation, local equilibrium is reached well after the appearance of nuclei with sharp interfaces, through a process of composition relaxation of the nuclei that is concomitant to their temporal growth. In addition, in liquid-liquid phase separation capillary forces are balanced by viscous forces, leading to a linear growth law of the nuclei sizes, while for liquid-vapor phase transition capillary forces are balanced by inertial forces, leading to a $\frac{2}{3}$ power-law behavior.

9 Conclusions

The main message of this review article is that we need to include intermolecular interactions in the classical hydrodynamic theory whenever we are interested in phenomena whose length scale is comparable to the interface thickness. This happens in the theory of contact line motion, the fluid mechanics of microdevices, in addition to the mixing and demixing of partially miscible mixtures that have been outlined here. In all cases, the final objective is that of determining, through the "correct" microscopic description, the appropriate boundary conditions of the classical equations of fluid dynamics that are applicable to macroscopic domains. The diffuse interface method, in particular, provides a sound theoretical basis for studying phase separating systems. Here, we have reviewed its basic theoretical foundations for both pure fluids and binary mixtures, above all when both phases behave like van der Waals fluids. We saw that the basic difference between the diffuse interface formulation and the classical approach is the presence of a capillary, or Korteweg, stress tensor in the momentum balance, expressing the tendency of the system to minimize its free energy. This extra stress gives rise to a body force, which is proportional to the gradient of the chemical potential difference, inducing a convection that, during phase transition, is much larger than that due to pure molecular diffusion. As it is identically zero at equilibrium, this force can be thought of as a non-equilibrium capillary force, whose net effect is that of generating an attractive force between domains of the same phase. Accordingly, coalescence is greatly enhanced during phase separation as the viscosity of the system decreases, i.e. when convection-induced fluxes become dominant over diffusion fluxes. This phenomenon is observed during the mixing process of liquid binary mixtures, where, however, once larger domains are formed, they must eventually dissolve by diffusion and, therefore, the process is actually retarded as the viscosity of the system decreases. Also, very interesting effects are observed when heat transfer is considered, as the potential gradient-induced convection arising during phase separation greatly accelerates the transport of heat. The influence of the such convection is clearly manifested during the phase separation of liquid binary mixtures, where the typical growth law for the domain size R changes from $R \propto t^{1/3}$, for very viscous systems, where diffusion prevails, to $R \propto t$, for convection-driven processes, as one can easily obtain by imposing that capillary forces are balanced either by diffusion or by viscous forces. In the case of single-component, liquid-vapor phase separation, instead, we obtain a $\frac{2}{3}$ power-law behavior, as capillary forces are balanced by inertial forces. In addition, here first the system reaches local equilibrium (with the formation of nuclei having sharp inter faces), and then these nuclei start to grow, while for viscous liquid binary mixtures the two events occur simultaneously.

Bibliography

D.M. Anderson, G.B. McFadden, and A.A. Wheeler. Diffuse-interface methods in fluid mechanics. *Annual Review of Fluid Mechanics*, 30:139–165, 1998.

L.K. Antanovskii. A phase field model of capillarity. *Physics of Fluids*, 7: 747–753, 1995.

L.K. Antanovskii. Microscale theory of surface tension. *Physical Review E*, 54:6285–6290, 1996.

D. Beysens, Y. Garrabos, V. S. Nikolayev, C. Lecoutre-Chabot, J.-P. Delville, and J. Hegseth. Liquid-vapor phase separation in a thermocapillary force field. *Europhysics Letters*, 59(2):245–251, 2002.

R. Borcia and M. Bestehorn. Phase-field simulations for drops and bubbles. *Physical Review E*, 75:056309, 2007.

J.W. Cahn. On spinodal decomposition. *Acta Metallurgica*, 9:795–801, 1961.

J.W. Cahn. Critical point wetting. *Journal of Chemical Physics*, 66(8): 3667–3772, 1977.

J.W. Cahn and J.E. Hilliard. Free energy of a nonuniform system. I. Interfacial free energy. *Journal of Chemical Physics*, 28:258–267, 1958.

J.W. Cahn and J.E. Hilliard. Free energy of a nonuniform system. III. Nucleation in a two-component incompressible fluid. *Journal of Chemical Physics*, 31:688–699, 1959.

F. Califano and R. Mauri. Drop size evolution during the phase separation of liquid mixtures. *Industrial & Engineering Chemistry Research*, 43: 349–353, 2004.

E.L. Cussler. *Diffusion*. Cambridge University Press, 1982.

H.T. Davis and L.E. Scriven. Stress and structure in fluid interfaces. *Advances in Chemical Physics*, 49:357–454, 1982.

P.G. de Gennes. Dynamics of fluctuations and spinodal decomposition in polymer blends. *Journal of Chemical Physics*, 72:4756–4763, 1980.

S.R. de Groot and P. Mazur. *Non-Equilibrium Thermodynamics*. Dover, New York, 1984.

B. U. Felderhof. Dynamics of the diffuse gas-liquid interface near the critical point. *Physica*, 48:541–560, 1970.

H. Furukawa. Role of inertia in the late stage of the phase separation of a fluid. *Physica A*, 204:237–245, 1994.

J.W. Gibbs. On the equilibrium of heterogeneous substances. *Transactions of the Connecticut Academy of Arts and Sciences*, 1876.

J.D. Gunton. Homogeneous nucleation. *Journal of Statistical Physics*, 95: 903–923, 1999.

R. Gupta, R. Mauri, and R. Shinnar. Liquid-liquid extraction using the composition induced phase separation process. *Industrial & Engineering Chemistry Research*, 35:2360–2368, 1996.

R. Gupta, R. Mauri, and R. Shinnar. Phase separation of liquid mixtures in the presence of surfactants. *Industrial & Engineering Chemistry Research*, 38:2418–2424, 1999.

P. C. Hohenberg and B. I. Halperin. Theory of dynamic critical phenomena. *Reviews of Modern Physics*, 49:435–479, 1977.

J.H. Israelachvili. *Intermolecular and Surface Forces.* Academic Press, 1992.

D. Jacqmin. Contact-line dynamics of a diffuse fluid interface. *Journal of Fluid Mechanics*, 402:57, 2000.

D. Jasnow and J. Viñals. Coarse-grained description of thermo-capillary flow. *Physics of Fluids*, 8:660–669, 1996.

K. Kawasaki. Kinetic equations and time correlation functions of critical fluctuations. *Annals of Physics*, 61:1–56, 1970.

D.J. Korteweg. Sur la forme que prennent les équations du mouvements des fluides si l'on tient compte des forces capillaires causées par des variations de densité considérables mais continues et sur la théorie de la capillarité dans l'hypothèse d'une variation continue de la densité. *Archives Néerlandaises des Sciences Exactes et Naturelles. Series II*, 6: 1–24, 1901.

A.G. Lamorgese and R. Mauri. Phase separation of liquid mixtures. In G. Continillo, S. Crescitelli, and M. Giona, editors, *Nonlinear Dynamics and Control in Process Engineering: Recent Advances*, pages 139–152. Springer, 2002.

A.G. Lamorgese and R. Mauri. Nucleation and spinodal decomposition of liquid mixtures. *Physics of Fluids*, 17:034107, 2005.

A.G. Lamorgese and R. Mauri. Mixing of macroscopically quiescent liquid mixtures. *Physics of Fluids*, 18:044107, 2006.

A.G. Lamorgese and R. Mauri. Diffuse-interface modeling of phase segregation in liquid mixtures. *International Journal of Multiphase Flow*, 34: 987–995, 2008.

A.G. Lamorgese and R. Mauri. Diffuse-interface modeling of liquid-vapor phase separation in a van der Waals fluid. *Physics of Fluids*, 21:044107, 2009.

L.D. Landau and E.M. Lifshitz. *Statistical Physics, Part I.* Pergamon Press, 1980.

M. Le Bellac. *Quantum and Statistical Field Theory.* Clarendon Press, 1991.

S. K. Lele. Compact finite-difference schemes with spectral-like resolution. *Journal of Computational Physics*, 103:16–42, 1992.

J. Lowengrub and L. Truskinovsky. Quasi-incompressible Cahn-Hilliard fluids and topological transitions. *Proceedings of the Royal Society of London, Series A*, 454:2617–2654, 1998.

T.C. Lucretius. *De Rerum Natura, Book I.* 50 B.C.E. "Corpus inani distinctum, quoniam nec plenum naviter extat nec porro vacuum." This is equivalent to one of the most basic principles of taoism, stating that nothing can be completely yin nor completely yang.

S. Madruga and U. Thiele. Decomposition driven interface evolution for layers of binary mixtures: II. Influence of convective transport on linear stability. *Physics of Fluids*, 21:062104, 2009.

R. Mauri, R. Shinnar, and G. Triantafyllou. Spinodal decomposition in binary mixtures. *Physical Review E*, 53:2613–2623, 1996.

D. Molin and R. Mauri. Enhanced heat transport during phase separation of liquid binary mixtures. *Physics of Fluids*, 19:074102, 2007.

D. Molin, R. Mauri, and V. Tricoli. Experimental evidence of the motion of a single out-of-equilibrium drop. *Langmuir*, 23:7459–7461, 2007.

S. Nagarajan, S. K. Lele, and J. H. Ferziger. A robust high-order compact method for large-eddy simulation. *Journal of Computational Physics*, 191:392–419, 2003.

S. Nagarajan, S. K. Lele, and J. H. Ferziger. Leading-edge effects in bypass transition. *Journal of Fluid Mechanics*, 572:471–504, 2007.

E.B. Nauman and D.Q. He. Nonlinear diffusion and phase separation. *Chemical Engineering Science*, 56:1999–2018, 2001.

A. Onuki. Dynamic van der Waals theory. *Physical Review E*, 75:036304, 2007.

A. Oprisan, S. A. Oprisan, J. Hegseth, Y. Garrabos, C. Lecoutre-Chabot, and D. Beysens. Universality in early-stage growth of phase-separating domains near the critical point. *Physical Review E*, 77(5):051118, 2008.

L.M. Pismen. Nonlocal diffuse interface theory of thin films and moving contact line. *Physical Review E*, 64:021603, 2001.

L.M. Pismen and Y. Pomeau. Disjoining potential and spreading of thin liquid layers in the diffuse-interface model coupled to hydrodynamics. *Physical Review E*, 62:2480–2492, 2000.

P. Poesio, G. Cominardi, A.M. Lezzi, R. Mauri, and G.P. Beretta. Effects of quenching rate and viscosity on spinodal decomposition. *Physical Review E*, 74:011507, 2006.

P. Poesio, A.M. Lezzi, and G.P. Beretta. Evidence of convective heat transfer enhancement induced by spinodal decomposition. *Physical Review E*, 75:066306, 2007.

P. Poesio, G.P. Beretta, and T. Thorsen. Dissolution of a liquid microdroplet in a nonideal liquid-liquid mixture far from thermodynamic equilibrium. *Physical Review Letters*, 103:064501, 2009.

S.D. Poisson. *Nouvelle Theorie de l'Action Capillaire.* Bachelier, 1831.

Lord Rayleigh. On the theory of surface forces. II. Compressible fluids. *Philosophical Magazine*, 33:209–220, 1892.

J.S. Rowlinson and B. Widom. *Molecular Theory of Capillarity.* Oxford University Press, 1982.

I. S. Sandler. *Chemical and Engineering Thermodynamics, 3rd Ed.* Wiley, 1999. Ch. 7.

G. M. Santonicola, R. Mauri, and R. Shinnar. Phase separation of initially non-homogeneous liquid mixtures. *Industrial & Engineering Chemistry Research*, 40:2004–2010, 2001.

E. Siggia. Late stages of spinodal decomposition in binary mixtures. *Physical Review A*, 20:595–605, 1979.

H. Tanaka. Coarsening mechanisms of droplet spinodal decomposition in binary fluid mixtures. *Journal of Chemical Physics*, 105:10099–10114, 1996.

H. Tanaka and T. Araki. Spontaneous double phase separation induced by rapid hydrodynamic coarsening in two-dimensional fluid mixtures. *Physical Review Letters*, 81:389–392, 1998.

U. Thiele, S. Madruga, and L. Frastia. Decomposition driven interface evolution for layers of binary mixtures: I. Model derivation and stratified base states. *Physics of Fluids*, 19:122106, 2007.

J.D. van der Waals. The thermodynamic theory of capillarity under the hypothesis of a continuous variation of density, 1893. Reprinted in *Journal of Statistical Physics,* 20:200–244 (1979).

N. Vladimirova and R. Mauri. Mixing of viscous liquid mixtures. *Chemical Engineering Science*, 59:2065–2069, 2004.

N. Vladimirova, A. Malagoli, and R. Mauri. Diffusion-driven phase separation of deeply quenched mixtures. *Physical Review E*, 58:7691–7699, 1998.

N. Vladimirova, A. Malagoli, and R. Mauri. Diffusio-phoresis of two-dimensional liquid droplets in a phase separating system. *Physical Review E*, 60:2037–2044, 1999a.

N. Vladimirova, A. Malagoli, and R. Mauri. Two-dimensional model of phase segregation in liquid binary mixtures. *Physical Review E*, 60:6968–6977, 1999b.

N. Vladimirova, A. Malagoli, and R. Mauri. Two-dimensional model of phase segregation in liquid binary mixtures with an initial concentration gradient. *Chemical Engineering Science*, 55:6109–6118, 2000.

G.W.F. von Leibnitz. *Nouveaux Essais sur l'Entendement Humain, Book II, Ch. IV.* 1765. Here Leibnitz applied to the natural world the statement "Natura non facit saltus" that in 1751 Linnaeus (i.e. Carl von Linné) in Philosophia Botanica, Ch. 77, had referred to species evolution.

B. Widom. Theory of phase equilibrium. *Journal of Physical Chemistry*, 100:13190–13199, 1996.

R. Yamamoto and K. Nakanishi. Computer simulation of vapor-liquid phase separation. *Molecular Simulation*, 16:119–126, 1996.

Phase separation of viscous ternary liquid mixtures

Jang Min Park * Roberto Mauri † and Patrick D. Anderson *

* Department Mechanical Engineering, Eindhoven University of Technology, 5600 MB Eindhoven, The Netherlands.

† Department of Chemical Engineering, Industrial Chemistry and Material Science, University of Pisa, 56126 Pisa, Italy.

Abstract In this work we study the demixing of ternary liquid mixtures. Our theoretical model follows the standard diffuse interface model, where convection and diffusion are coupled via a body force, expressing the tendency of the mixture to minimize its free energy. This driving force induces a material flux which, in most cases, is much larger than that due to pure molecular diffusion. In fact, here we model the behavior of a very viscous polymer melt, so that the Peclet number, expressing the ratio between convective and diffusive mass fluxes, is equal to 50. To simulate the system, the already existent TFEM code was extended to three component systems. Two examples are presented, with known free energy expressions, to describe the behavior of partially miscible three phase mixtures.

1 Introduction

The diffuse interface model was developed originally to describe near-critical behavior of single-component fluids and partially miscible binary mixtures (Cahn and Hilliard, 1958, 1959; Hohenberg and Halperin, 1977; Lowengrub and Truskinovsky, 1998; Vladimirova et al., 1999b) (see the review article by Lamorgese, Molin and Mauri presented in this book). Despite many industrial and biochemical processes involve mixtures with three or more components, only a few theoretical and numerical works have studied these systems. In particular, Huang et al. (1995) analyzed numerically the dynamics of phase separation of ternary alloys (i.e. where convective effects can be neglected) into two and three phases by solving the nonlinear spinodal decomposition equations in two dimensions. Examining the dynamical scaling and the growth laws for the late stages of separation, they saw that the growth law $R(t) \propto t^{1/3}$ is always obeyed, despite the fact that the self-similar regime is achieved very slowly in ternary systems. Later, Kim and Lowengrub (2004, 2005) developed a full NS-CH code to model the phase mixing / demixing and the Rayleigh instability of ternary mixtures using a diffuse interface model in the low Reynolds number regime. There, applying boundary integral methods, the effects of surfactants on drop dynamics, tip-streaming and drop deformation have been investigated. Phase-field ternary mixture models have also been of interest for modeling many different physical phenomena, from solidification and microstructure evolution in ternary alloy systems Kobayashi et al. (2003) to mutual diffusion effects in partially miscible polymer blends Tufano et al. (2010), to surfactant-induced emulsion coarsening (Lamorgese and Banerjee, 2011).

In this work, starting from the already existent results for two component systems (Lamorgese and Mauri, 2005), we develop a general model of ternary mixtures, in which the Navier-Stokes equation is coupled to generalized Cahn-Hilliard equations for the phase variables. Compared to previous models, the present one has the advantage of simplicity and thermodynamic consistency, without employing any *ad-hoc* term that cannot be directly related to macroscopic, easily measured parameters.

2 The governing equations

2.1 Multi-component mixtures at equilibrium

Consider a homogeneous mixture of N species A_k $(k = 1, 2 \ldots N)$, with molar fractions x_k, kept at temperature T and pressure P. For sake of simplicity, in our model we assume that the molecular weights, specific volumes and viscosities of all species are the same, namely $M_k = M_w$, $\bar{V}_k = \bar{V}$ and $\eta_k = \eta$, for all species k, so that molar, volumetric and mass fractions are all equal to each other, and the mixture viscosity is composition-independent. The equilibrium state of this system is described by the "coarse-grained" free energy functional, that is the molar Gibbs energy of mixing, Δg^{th},

$$\Delta g^{th} = g^{th} - \sum_{k=1}^{N} g_k x_k, \tag{1}$$

where g^{th} is the energy of the mixture at equilibrium, while g_k is the molar free energy of pure species A_k at temperature T and pressure P. The free energy Δg^{th} is the sum of an ideal part Δg^{id} and a so-called excess part g^{ex}, with

$$\Delta g^{id} = RT \sum_{k=1}^{N} x_k \log x_k, \tag{2}$$

where R is the gas constant, while the excess molar free energy can be expressed as,

$$g^{ex} = \frac{1}{2} RT \sum_{i,k=1}^{N} \Psi_{ik} x_i x_k, \tag{3}$$

where Ψ_{ik} are functions of T and P, with $\Psi_{ik} = \Psi_{ki}$ and $\Psi_{ii} = 0$. This expression can be generally derived by considering the molecular interactions between nearest neighbors or summing all pairwise interactions throughout the whole system (Lifshitz and Pitaevskii, 1984). As shown by Mauri et al. (1996), Eq. (3) it can also be derived from first principles, assuming that the pairwise forces between identical molecules, $F_{i,i}$ are all equal to each other and larger than the pairwise forces among unequal molecules, F_{ij} (with $i \neq j$), i.e. $F_{i,i} = F_{j,j} > F_{i,j}$, obtaining an expression for Ψ_{ij} which depends on $(F_{i,i} - F_{i,j})$. In the following, we shall assume that P is fixed, so that the physical state of the mixture at equilibrium depends only on T and x_i.

Now, it is well-known that any variation of the molar free energy can be written as (Prausnitz et al., 1986),

$$dg^{th} = RT \sum_{i=1}^{N} \mu_i^{th} dx_i, \tag{4}$$

where μ_i^{th} denotes the chemical potential of species A_i in solution, i.e.,

$$\mu_i^{th} = \frac{1}{RT} \frac{\partial (c g^{th})}{\partial c_i}, \tag{5}$$

with $c_i = c x_i$ denoting the mole densities, that is the number of moles per unit volume, of species A_i, and $c = \sum c_i$ is the total mole density. In our case we obtain:

$$\mu_i^{th} = \frac{g^{th}}{RT} + \log x_i + \sum_{k=1}^{N} \Psi_{ik} (1 - x_i) x_k - \sum_{j,k \neq i=1}^{N} \Psi_{jk} x_j x_k. \tag{6}$$

Since free energy is an extensive quantity, it is easy to show that chemical potentials represent the amount of free energy due to each species, i.e.

$$g^{th} = RT \sum_{i=1}^{N} \mu_i^{th} x_i. \tag{7}$$

Therefore, comparing Eqs. (4) and (7), we obtain the Gibbs-Duhem relation,

$$\sum_{i=1}^{N} x_i d\mu_i^{th} = 0. \tag{8}$$

Considering that $\sum_{i=1}^{N} x_i = 1$, we see that Eq. (4) can be rewritten as:

$$dg^{th} = RT \sum_{i=1}^{N-1} \mu_{iN}^{th} dx_i, \tag{9}$$

where $\mu_{ij}^{th} \equiv \mu_i^{th} - \mu_j^{th}$. Accordingly, we see that the quantities x_i and $RT\mu_{iN}^{th}$ are thermodynamically conjugated, i.e. $RT\mu_{iN}^{th} = \partial g_{eq}/\partial x_i$. In fact, applying this expression, we obtain:

$$\mu_{ij}^{th} = \ln \frac{x_i}{x_j} + \Psi_{ij}(x_j - x_i) + \sum_{k \neq i,j=1}^{N} (\Psi_{ik} - \Psi_{jk}) x_k, \tag{10}$$

where, by definition, $\mu_{ij}^{th} \equiv \mu_{ik}^{th} - \mu_{jk}^{th}$ and $\mu_{ij}^{th} = -\mu_{ji}^{th}$. The same result could be obtained directly from Eq. (6).

2.2 Non local terms

In order to take into account the effects of spatial inhomogeneities, following Cahn and Hilliard (1958, 1959), we assume that the total, or generalized, free energy $\widetilde{g}$ is the sum of an equilibrium part and a non local part,

$$\widetilde{g} = g^{th} + g^{nl}, \tag{11}$$

where the latter is given by the following expression:

$$g^{nl} = \frac{1}{4} RTa^2 \sum_{i \neq j=1}^{N} (\nabla x_i)^2. \tag{12}$$

Here a represents the typical length of spatial inhomogeneities in the composition which, as shown by van der Waals (1893), is proportional to the surface tension between the two phases. Note that, considering that $\sum_{i=1}^{N} x_i = 1$, this expression can also be written as:

$$g^{nl} = -\frac{1}{2} RTa^2 \sum_{i \neq j=1}^{N} \nabla x_i \nabla x_j = \frac{1}{2} RTa^2 \sum_{i=1}^{N-1} \left(\nabla x_i \sum_{j \geq i}^{N-1} \nabla x_j \right). \tag{13}$$

Now, chemical potentials can be generalized as follows:

$$\widetilde{\mu}_i = \frac{1}{RT} \frac{\delta (c\widetilde{g})}{\delta c_i} = \mu_i^{th} + \mu_i^{nl}, \tag{14}$$

where μ_i^{th} is defined in Eq. (5), while,

$$\mu_i^{nl} = -\frac{1}{RT}\nabla\cdot\left(\frac{\partial(cg^{nl})}{\partial\nabla c_i}\right). \tag{15}$$

Consequently we obtain:

$$\mu_i^{nl} = \frac{a^2}{2}\nabla\cdot\left[-(1-x_i)\nabla x_i + \sum_{j\neq i=1}^{N} x_j\nabla x_j\right]. \tag{16}$$

It can be shown that, being g^{nl} a quadratic function of ∇x_i, we obtain:

$$2g^{nl} = \sum_{i=1}^{N}\mu_i^{nl}x_i. \tag{17}$$

In addition,

$$-\nabla g^{nl} = \sum_{i=1}^{N}\mu_i^{nl}\nabla x_i. \tag{18}$$

Consequently,

$$\sum_{i=1}^{N} x_i\nabla\mu_i^{nl} \neq 0, \tag{19}$$

showing that the Gibbs-Duhem relation cannot be extended to the non local part of the free energy.

Finally, since, as we saw, we are mostly interested in the chemical potential differences, we have:

$$\mu_{ij}^{nl} = -\frac{a^2}{2}\nabla^2(x_i - x_j). \tag{20}$$

In particular, for ternary mixtures, we obtain:

$$\widetilde{\mu}_{12} = \ln\frac{x_1}{x_2} + \Psi_{12}(x_2 - x_1) + (\Psi_{13} - \Psi_{23})x_3 - \frac{a^2}{2}\nabla^2(x_1 - x_2), \tag{21}$$

$$\widetilde{\mu}_{23} = \ln\frac{x_2}{x_3} + \Psi_{23}(x_3 - x_2) + (\Psi_{21} - \Psi_{31})x_1 - \frac{a^2}{2}\nabla^2(x_2 - x_3), \tag{22}$$

$$\widetilde{\mu}_{31} = \ln\frac{x_3}{x_1} + \Psi_{31}(x_1 - x_3) + (\Psi_{32} - \Psi_{12})x_2 - \frac{a^2}{2}\nabla^2(x_3 - x_1). \tag{23}$$

Note that, by definition, $\widetilde{\mu}_{12} \equiv \widetilde{\mu}_{13} + \widetilde{\mu}_{32}$ and $\widetilde{\mu}_{ij} = -\widetilde{\mu}_{ji}$.

2.3 The equations of motion

Consider a mixture of N components, where the k-th species has density ρ_k, volume fraction ϕ_k and velocity $\mathbf{v}_k$. The mass balance equation for each component can be written as:

$$\frac{\partial(\rho_k \phi_k)}{\partial t} + \nabla \cdot (\rho_k \phi_k \mathbf{v}_k) = 0. \tag{24}$$

Now, define the mixture density and the mixture velocity as the following mass-average quantities,

$$\rho = \sum_{k=1}^{N_f} \rho_k \phi_k. \tag{25}$$

and

$$\mathbf{v} = \frac{1}{\rho} \sum_{k=1}^{N_f} \phi_k \rho_k \mathbf{v}_k = \sum_{k=1}^{N_f} c_k \mathbf{v}_k, \tag{26}$$

where $c_k = \phi_k \rho_k / \rho$ is the mass fraction of the k-th component. Summing equation (24) for all k's, we trivially obtain the continuity equation,

$$\frac{\partial \rho}{\partial t} + \nabla \cdot (\rho \mathbf{v}) = 0. \tag{27}$$

In the following, we assume that all components have the same density ρ and molecular weight M_w, so that $c_k = \phi_k = x_k$ represent the mass, volume and molar fractions of the k-th component. Accordingly, the mixture is incompressible, i.e. Eq. (27) reduces to

$$\nabla \cdot \mathbf{v} = 0, \tag{28}$$

while the mass concentration equations becomes:

$$\dot{x}_k = \frac{\partial x_k}{\partial t} + \mathbf{v} \cdot \nabla x_k = -\nabla \cdot \mathbf{J}_k, \tag{29}$$

where the dot indicates the advection derivative with respect to the mixture velocity $\mathbf{v}$ and

$$\mathbf{J}_k = x_k (\mathbf{v}_k - \mathbf{v}) \tag{30}$$

is the volumetric diffusive flux, depending on the the velocity of the k-th component with respect to the mean, with $\sum \mathbf{J}_k = \mathbf{0}$. Obviously, since $\sum c_k = 1$, only $N-1$ of these equations are independent.

Equations (28) and (29) must be couple with the Navier-Stokes equation,

$$\rho \dot{\mathbf{v}} + \nabla p = \nabla \cdot \boldsymbol{\tau} + \mathbf{f}_\phi, \tag{31}$$

where $\boldsymbol{\tau}$ is the viscous stress tensor, while $\mathbf{f}_\phi$ is the Korteweg reversible force,

$$\mathbf{f}_\phi = \frac{\rho RT}{M_w}\frac{\delta \widetilde{g}}{\delta \mathbf{r}} = \frac{\rho RT}{M_w}\sum_{i=1}^{N}\left(\widetilde{\mu}_i \nabla x_i\right). \tag{32}$$

Note that, since $\widetilde{\mu}_i = \mu_i^{th} + \mu_i^{nl}$ and considering that

$$\sum \mu_i^{th} \nabla x_i = \nabla g^{th}, \tag{33}$$

after redefining the pressure as $p - \rho RT g^{th}$, the above equation can be rewritten replacing $\widetilde{\mu}_i$ with μ_i^{nl}. Now, considering that $\sum x_i = 1$, we finally obtain:

$$\mathbf{f}_\phi = \frac{\rho RT}{M_w}\sum_{i=1}^{N-1}\left(\mu_{iN}^{nl} \nabla x_i\right). \tag{34}$$

2.4 The constitutive equations for the diffusive fluxes

Here, for sake of simplicity, we assume that the viscosities of all components are the same, so that the viscous stress in Eq. (31) can be written:

$$\boldsymbol{\tau} = \eta\left(\nabla \mathbf{v} + (\nabla \mathbf{v})^{\dagger}\right), \tag{35}$$

where η is the uniform viscosity of the mixture.

As for the material diffusive fluxes, generalizing the two component case, we use the following constitutive relations,

$$\mathbf{J}_i = -\sum_{j=1}^{N} D_{ij} x_i x_j \nabla \widetilde{\mu}_{ij}, \tag{36}$$

here $D_{ij} = D_{ji}$ and $\mu_{ij} = \mu_i - \mu_j$, so that $\sum \mathbf{J}_i = \mathbf{0}$ identically, as it should.

Note that, had we assumed the following constitutive relation,

$$\mathbf{J}_i = -D x_i \nabla \mu_i, \tag{37}$$

the sum of the diffusive fluxes $\mathbf{J}_i$ would not be zero, as the Gibbs-Duhem relation is not satisfied by the non local part of the chemical potential.

In particular, for ternary mixtures, assuming that the diffusion coefficients are all equal, we obtain:

$$\mathbf{J}_1 = -D x_1 x_2 \nabla \widetilde{\mu}_{23} - D x_1 (1 - x_1) \nabla \widetilde{\mu}_{13}, \tag{38}$$

$$\mathbf{J}_2 = -D x_1 x_2 \nabla \widetilde{\mu}_{13} - D x_2 (1 - x_2) \nabla \widetilde{\mu}_{23}. \tag{39}$$

Note that this makes the concentration equation a fourth order nonlinear advection diffusion equation, which is a generalization of the classical Cahn-Hilliard equation, used to describe the phase separation of binary mixtures.

The above equations can be re-scaled defining:

$$\widehat{\mathbf{r}} = \frac{\mathbf{r}}{a}; \quad \widehat{t} = \frac{t}{(a^2/D)}; \quad \widehat{\mathbf{v}} = \frac{\mathbf{v}}{V}; \quad \widehat{p} = \frac{p}{\eta V/a}, \tag{40}$$

where V is a characteristic velocity, which can be estimated through Eqs. (31) as $V \approx f_\phi a^2/\eta$, with $f_\phi \approx \rho RT/(aM_w)$. At the end, we obtain the following non-dimensional system of equations, which is the ternary version of the model *H*, in the nomenclature of Hohenberg and Halperin **?**,

$$\nabla p = \nabla^2 \mathbf{v} + (\widetilde{\mu}_{13} \nabla x_1 + \widetilde{\mu}_{23} \nabla x_2); \tag{41}$$

$$\nabla \cdot \mathbf{v} = 0; \tag{42}$$

$$\frac{\partial x_1}{\partial t} + N_{Pe} \mathbf{v} \cdot \nabla x_1 = \nabla \cdot \left[x_1 x_2 \nabla \widetilde{\mu}_{23} + x_1 (1 - x_1) \nabla \widetilde{\mu}_{13} \right]; \tag{43}$$

$$\frac{\partial x_2}{\partial t} + N_{Pe} \mathbf{v} \cdot \nabla x_2 = \nabla \cdot \left[x_1 x_2 \nabla \widetilde{\mu}_{13} + x_2 (1 - x_2) \nabla \widetilde{\mu}_{23} \right], \tag{44}$$

where hats have been omitted for simplicity, while $\widetilde{\mu}_{13}$ and $\widetilde{\mu}_{23}$ are given by Eq. (22) and (23). Here, $N_{Pe} = Va/D$ defines the capillary Peclet number, denoting the ratio between convective and dilusive mass fluxes,

$$N_{Pe} = \frac{a^2}{D} \frac{\rho}{\eta} \frac{RT}{M_w}. \tag{45}$$

For common liquids, N_{Pe} is very large, showing that diffusion is important only in the vicinity of local equilibrium, when the body force f_ϕ is negligibly small. Here, we assume that $N_{Pe} = 50$, corresponding to a typical value of polymer melts. In fact, reasonable numerical values in this case are: $a \approx 10^{-5} cm$, $D \approx 10^{-7} cm^2/s$, $\nu \approx 10^2 cm^2/s$, $M_w \approx 10^4 g/gmole$.

3 Numerical results

Physically, we consider an instantaneous quench bringing the mixture from its single-phase, stable and homogeneous initial state to an unstable final state, corresponding to a point in its phase diagram lying within the coexistence curve.

Two types of binary mixtures were considered, corresponding to mixtures that phase separate in two and in three coexisting phases. In the following, these two cases will be treated separately.

3.1 Numerical methods

The two second-order differential equations representing the chemical potential and the concentration equation are solved in a coupled way for both phases. For the temporal discretization a first-order Euler implicit scheme was employed, with a 4×10^{-3} non-dimensional time step. The nonlinear term in the chemical potential equation was linearized by a standard Picard method. A second-order finite element method was used for spatial discretization of the set of equations, using a square periodic domain, with 40×40 elements, assuming periodic boundary conditions.

Details about the iteration scheme can be found in Keestra et al. (2003) and Khatavkar et al. (2006). The flow problem was solved using the velocity-pressure formulation and discretized by a standard Garlekin finite element method. Taylor-Hood quadrilateral elements with continuous pressure that employ a biquadratic approximation for the velocity and a bilinear approximation for the pressure are used. The resulting discretized second-order linear algebraic equation was solved using a direct method based on a sparse multifrontal variant of Gaussian elimination (HSL/MA41). (Amestoy and Duff; Amestoy and Puglisi)

3.2 Two-phase mixture.

Assume that $\Psi_{12} = 4$ and $\Psi_{13} = \Psi_{23} = 0$, corresponding to the triangular phase diagram of Fig. 1. Here, the line $1 - 2$ represents an almost immiscible binary mixture, with equilibrium points having composition $(x_1; x_2) = (x; 1 - x)$ and $(1 - x; x)$, with $x = 0.021$ (we find that from Eq. (21), imposing that $\mu_{12} = 0$, with x_1 and x_2 uniform, so that there is no contribution from the non-local part; the additional solution, $x = 1/2$, is unstable.) As we add the 3-component, which is miscible with both 1 and 2, the mixture becomes more and more miscible, until we reach the critical point C at $(x_1 = x_2 = 0.224; x_3 = 0.552)$ (we find that imposing that at the critical point $x_1 = x_2 = x$, so that $x_3 = 1 - 2x$, and then imposing that the chemical potential differences are all zero.) Note that because of symmetry all tye lines are parallel to $1 - 2$ axis.

Initially, the mixture is assumed to be homogeneous with $x_1 = x_2 = 0.35$ and $x_3 = 0.3$, corresponding to a state well inside the unstable region of the phase diagram. When a small random perturbation is superimposed, as shown in Fig. 2-4, we see that the mixture separates into two phases, one rich of component 1 and the other rich of component 2, while component 3, being equally miscible within 1 and 2, distributes homogeneously within the mixture. As one should expect, the dynamics of phase separation in this case is identical to that for binary mixtures and therefore all the comments that were made for binary mixtures can also be applied here. In particular, we find that the typical size R of single phase domains grows with time as $R(t) \propto t^n$, where, as expected, the power-law scaling exponent n varies from 1/3 (diffusion controlled) to 1 (hydrodynamics controlled), corresponding to cases with $N_{Pe} = 0$ and $N_{Pe} \gg 1$, respectively (see Fig. 4 in Lamorgese and Mauri (2008)).

3.3 Three-phase mixture.

Consider a symmetric ternary mixture, $\Psi_{12} = \Psi_{13} = \Psi_{23} = \Psi = 4$. As shown in Fig. 5, at the center of the triangular phase diagram there in a smaller triangle (called tie triangle), whose three vertices represent the composition of three coexisting phases, α in point A, β in point B and γ in point C (Huang et al., 1995). If x is the distance between A and the $1-2$ axis (and between B and the $1-3$ axis as well as between C and the $2-3$ axis, by symmetry), the compositions of the three coexisting phases are:

$$\mathbf{x}^\alpha = (1-2x,\, x,\, x)\,; \quad \mathbf{x}^\beta = (x,\, 1-2x,\, x)\,; \quad \mathbf{x}^\gamma = (x,\, x,\, 1-2x)\,, \tag{46}$$

with $x < 1/3$. Then, from Eq. (21)-(23), we see that all thermodynamic chemical potential differences in A, B and C are identically zero, which gives us the following relation:

$$\mu_{12}^\alpha = \ln \frac{1-2x}{x} + \Psi(3x-1) = 0. \tag{47}$$

(all the other relations reduce to this one or are identically satisfied) So, assuming that we have three phases, knowing Ψ, we can find x, i.e. the composition of the three coexisting phases. Here, we see that the composition $x = 1/3$ always satisfies this expression. However, this correspond to a point of stable equilibrium only for $\Psi < 8/3$. When, as in our case, $\Psi > 8/3$, we easily find:

$$x = \frac{1}{4}\left(1 - \sqrt{1 - \frac{8}{3\Psi}}\right). \tag{48}$$

(We discard the other solution, as it does not satisfy the requirement $x < 1/3$.) Clearly, we see that when $\Psi = 8/3$ the three points A, B and C converge at center of the triangle, while when $\Psi \gg 8/3$ they end up at the three vertices 1, 2 and 3 of the phase diagram, as it should be. In our case, with $\Psi = 4$, we find $x = 0.394$, corresponding to $\mathbf{x}^\alpha = (0.212; 0.394; 0.394)$; $\mathbf{x}^\beta = (0.394; 0.212; 0.394)$; $\mathbf{x}^\gamma = (0.394; 0.394; 0.212)$.

Initially, the mixture is assumed to be homogeneous with $x_1 = x_2 = x_3 = 1/3$, corresponding again to an unstable state. When a small random

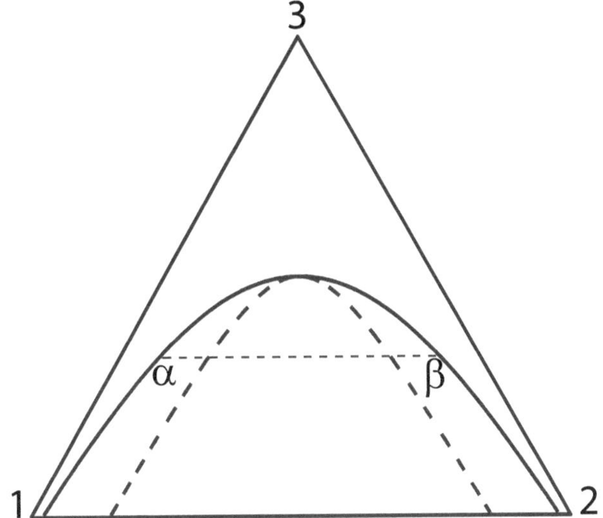

Figure 1. Phase diagram for $\Psi_{12} = 4$, $\Psi_{23} = \Psi_{31} = 0$.

perturbation is superimposed, as shown in Fig. 6-8, we see that the mixture separates into three coexisting phases. Unexpectedly, though, the perfect symmetry among the three components gets broken. In fact, while phase γ separates very quickly, as seen in Fig. 8, the other two phases separates later. In addition, the morphology of the system appear to be crystal-like, as if the different phases were forced to orient themselves at certain angles with respect to each other. Consequently, the growth rate of the single phase domains does not appear to be unique, although further work is required to give a definitive answer.

Bibliography

P.R. Amestoy and I.S. Duff. Vectorization of a multiprocessor multifrontal code. *International Journal of Supercomputer Applications*, page 41.

P.R. Amestoy and C. Puglisi. An unsymmetrized multifrontal lu factorization. *SIAM Journal on Matrix Analysis and Applications*, page 553.

J.W. Cahn and J.E. Hilliard. Free energy of a nonuniform system. I. Interfacial free energy. *Journal of Chemical Physics*, 28:258–267, 1958.

J.W. Cahn and J.E. Hilliard. Free energy of a nonuniform system. III. Nucleation in a two-component incompressible fluid. *Journal of Chemical Physics*, 31:688–699, 1959.

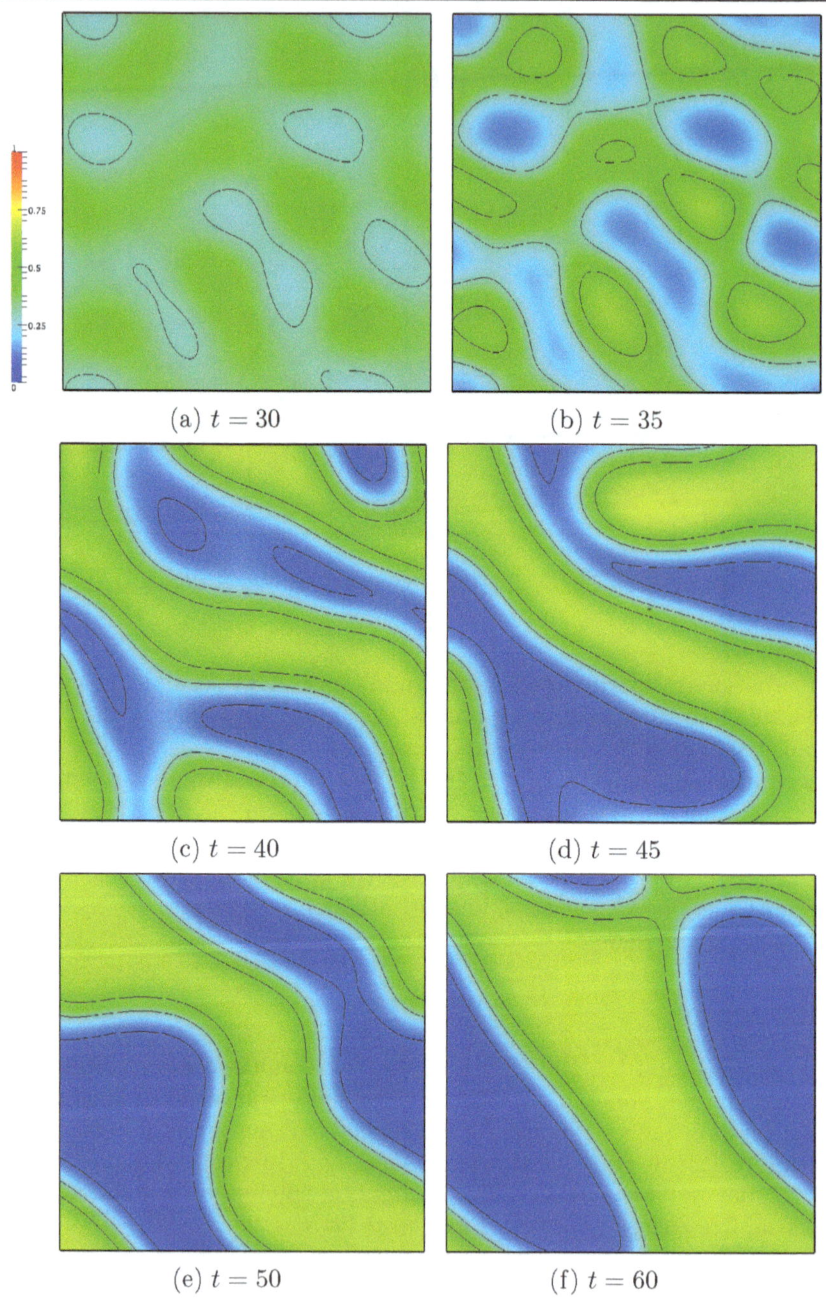

(a) $t = 30$ (b) $t = 35$

(c) $t = 40$ (d) $t = 45$

(e) $t = 50$ (f) $t = 60$

Figure 2. Evolution of x_1 for two-phase mixture. Contour curves are of x_1 = 0.1, 0.3, 0.5, 0.7 and 0.9.

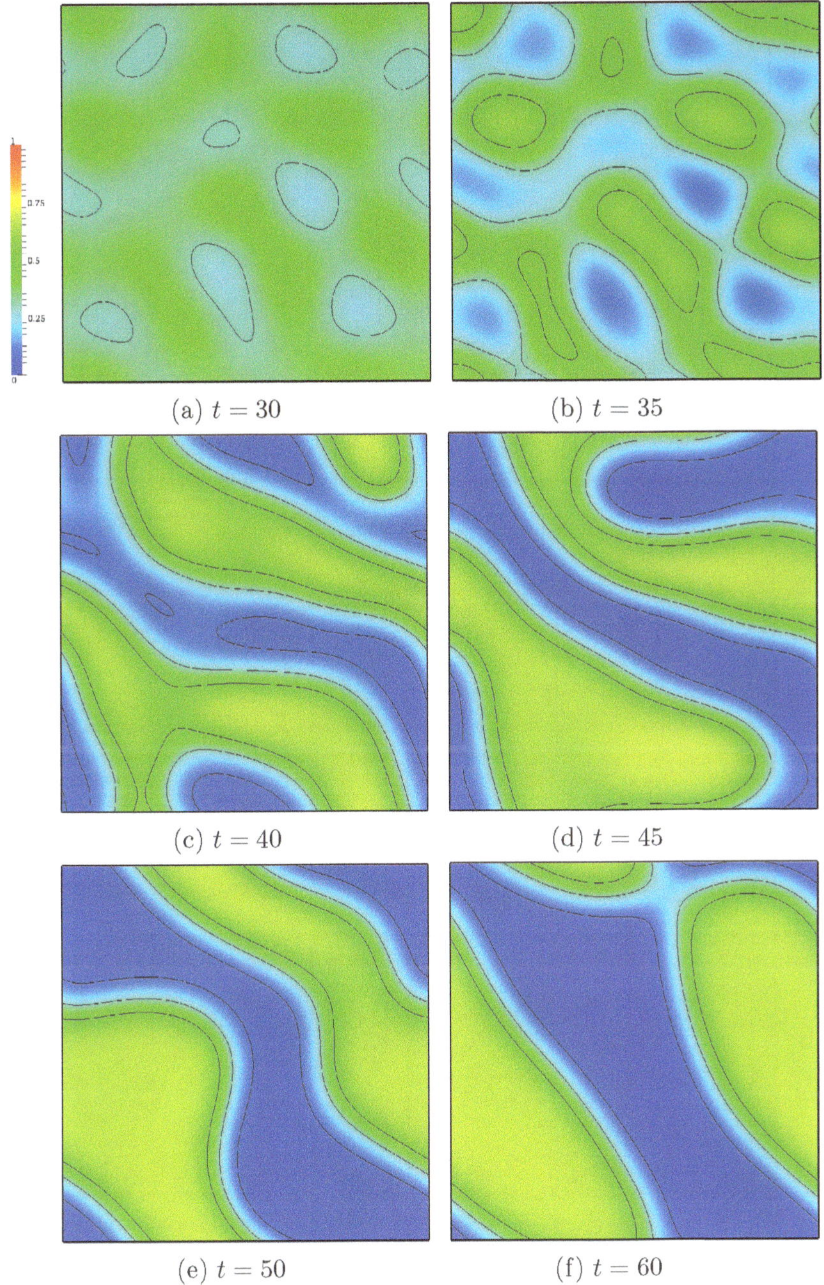

(a) $t = 30$ (b) $t = 35$

(c) $t = 40$ (d) $t = 45$

(e) $t = 50$ (f) $t = 60$

Figure 3. Evolution of x_2 for two-phase mixture. Contour curves are of x_2 = 0.1, 0.3, 0.5, 0.7 and 0.9.

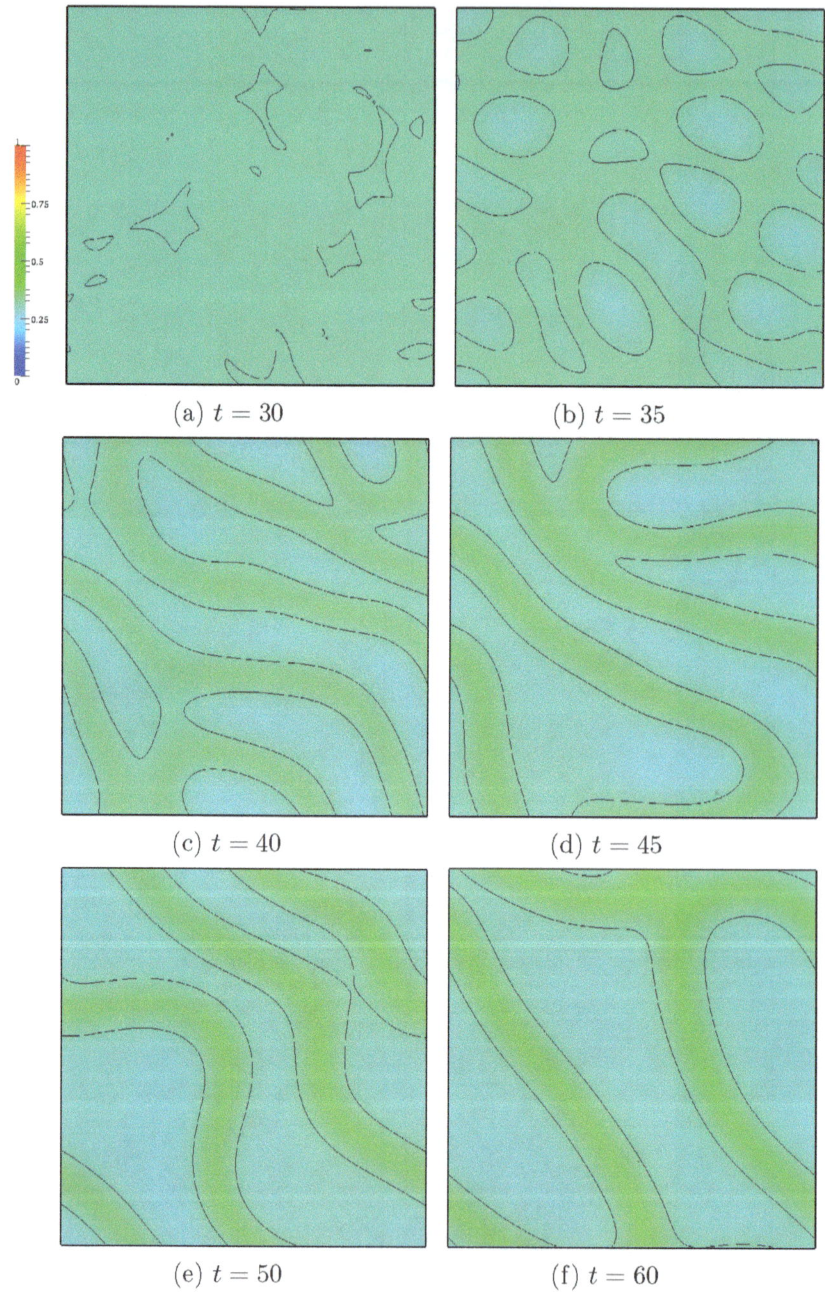

(a) $t = 30$ (b) $t = 35$

(c) $t = 40$ (d) $t = 45$

(e) $t = 50$ (f) $t = 60$

Figure 4. Evolution of x_3 for two-phase mixture. Contour curves are of x_3 = 0.1, 0.3, 0.5, 0.7 and 0.9.

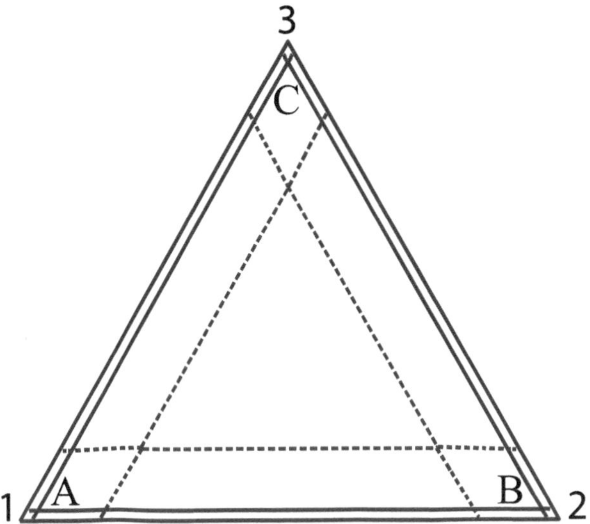

Figure 5. Phase diagram for $\Psi_{12} = \Psi_{23} = \Psi_{31} = 4$.

P. C. Hohenberg and B. I. Halperin. Theory of dynamic critical phenomena. *Reviews of Modern Physics*, 49:435–479, 1977.

C. Huang, M. Olvera de la Cruz, and B.W. Swift. Phase separation of ternary mixtures: symmetric polymer blends. *Macromolecules*, 28:7996, 1995.

B. Keestra, P.C.J. van Puyvelde, P.D. Anderson, and H.E.H. Meijer. Diffuse interface modeling of the morphology and rheology of immiscible polymer blends. *Physics of Fluids*, 15:2567–2575, 2003.

V.V. Khatavkar, P.D. Anderson, and H.E.H. Meijer. On scaling of diffuse-interface models. *Chemical Engineering Science*, 61:2364–2368, 2006.

J.S. Kim and J. Lowengrub. Conservative multigrid methods for ternary cahn-hilliard systems. *Communications in Mathematical Sciences*, 2:53, 2004.

J.S. Kim and J. Lowengrub. Phase field modeling and simulation of three phase flows. *Interfaces and Free Boundaries*, 7:435, 2005.

H. Kobayashi, M. Ode, S. G. Kim, W. T. Kim, and T. Suzuki. Phase-field model for solidification of ternary alloys coupled with thermodynamic database. *Scripta Materialia*, 48:689, 2003.

A.G. Lamorgese and S. Banerjee. Insoluble surfactant efefcts on emulsion coarsening in a gravitational field via phase-field ternary mixture model.

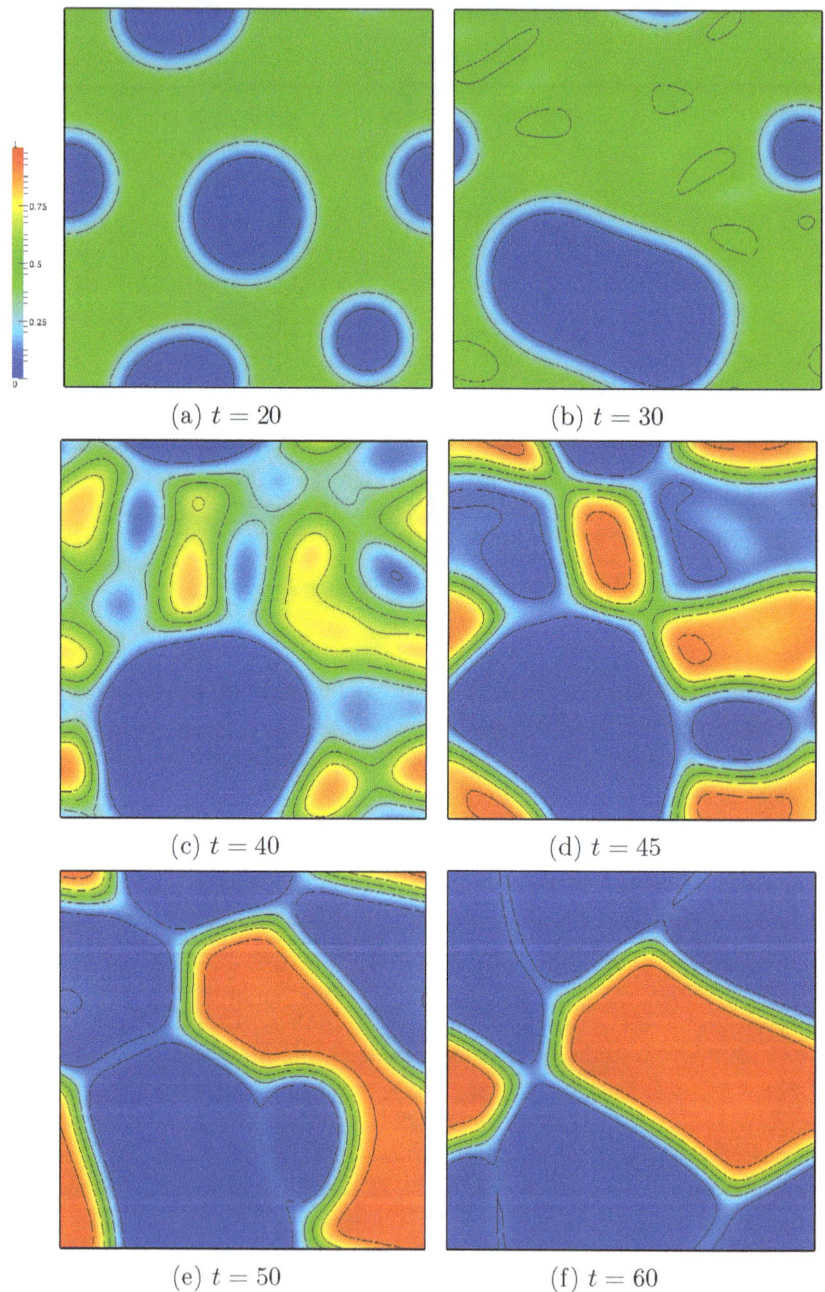

(a) $t = 20$ (b) $t = 30$

(c) $t = 40$ (d) $t = 45$

(e) $t = 50$ (f) $t = 60$

Figure 6. Evolution of x_1 for three-phase mixture. Contour curves are of $x_1 = 0.1, 0.3, 0.5, 0.7$ and 0.9.

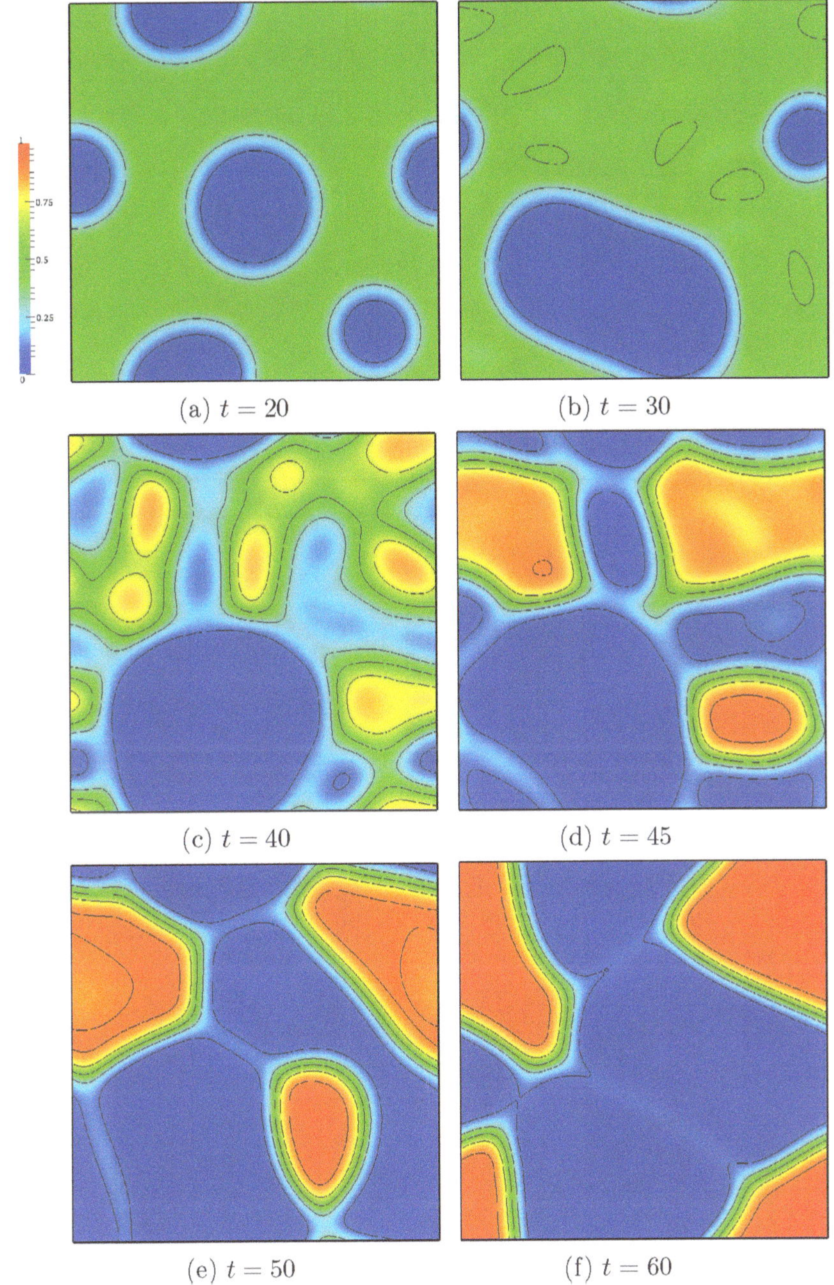

(a) $t = 20$ (b) $t = 30$

(c) $t = 40$ (d) $t = 45$

(e) $t = 50$ (f) $t = 60$

Figure 7. Evolution of x_2 for three-phase mixture. Contour curves are of $x_2 = 0.1, 0.3, 0.5, 0.7$ and 0.9.

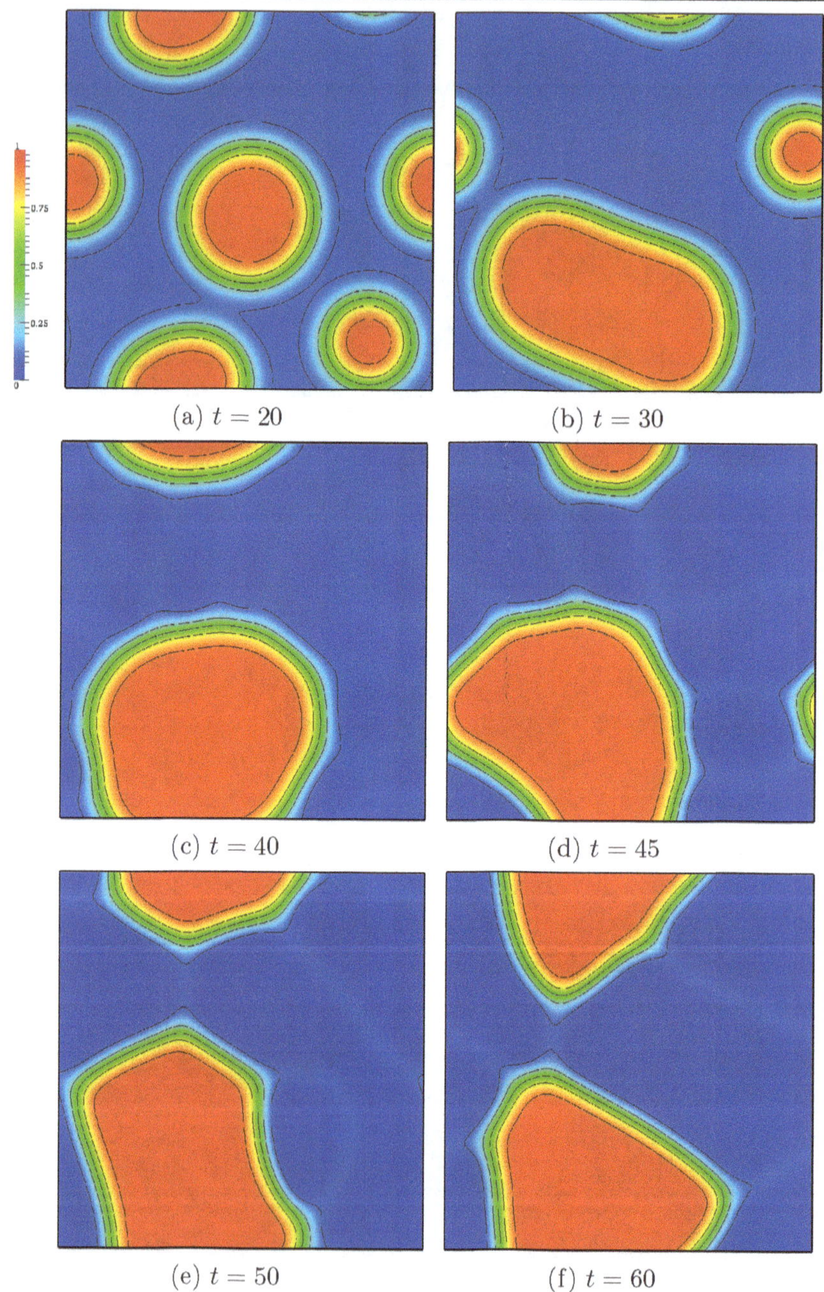

(a) $t = 20$ (b) $t = 30$

(c) $t = 40$ (d) $t = 45$

(e) $t = 50$ (f) $t = 60$

Figure 8. Evolution of x_3 for three-phase mixture. Contour curves are of $x_3 = 0.1$, 0.3, 0.5, 0.7 and 0.9.

In *Proceedings of the 8-th SINTEF/NTNU Conference*, Trondheim, Norway, 2011.

A.G. Lamorgese and R. Mauri. Nucleation and spinodal decomposition of liquid mixtures. *Physics of Fluids*, 17:034107, 2005.

A.G. Lamorgese and R. Mauri. Diffuse-interface modeling of phase segregation in liquid mixtures. *International Journal of Multiphase Flow*, 34: 987–995, 2008.

E. M. Lifshitz and L. P. Pitaevskii. *Physical Kinetics*. Pergamon Press, New York, 1984.

J. Lowengrub and L. Truskinovsky. Quasi-incompressible Cahn-Hilliard fluids and topological transitions. *Proceedings of the Royal Society of London, Series A*, 454:2617–2654, 1998.

R. Mauri, R. Shinnar, and G. Triantafyllou. Spinodal decomposition in binary mixtures. *Physical Review E*, 53:2613–2623, 1996.

J.M. Prausnitz, R.N. Lichtenthaler, and E. Gomes de Azevedo. *Molecular Thermodynamics of Fluid-hase Equilibria, 2nd Ed.* Prentice Hall, 1986.

C. Tufano, G. W. M. Peters, H. E. H. Meijer, and P. D. Anderson. Effects of partial miscibility on drop-wall and drop-drop interactions. *Jornal of Rheology*, 54:159, 2010.

J.D. van der Waals. The thermodynamic theory of capillarity under the hypothesis of a continuous variation of density, 1893. Reprinted in *Journal of Statistical Physics,* 20:200–244 (1979).

N. Vladimirova, A. Malagoli, and R. Mauri. Two-dimensional model of phase segregation in liquid binary mixtures. *Physical Review E*, 60:6968–6977, 1999b.

Dewetting and decomposing films of simple and complex liquids

Uwe Thiele
Department of Mathematical Sciences, Loughborough University, Loughborough, Leicestershire LE11 3TU, UK

Abstract
We provide a brief account of recent studies of dewetting films of simple and some complex liquids. First, we review basic models for dewetting one-layer films of simple liquids as they are often employed as reference case for studies of simple liquids in more complex situations or of complex liquids. Then we discuss films of binary mixtures that may undergo dewetting and decomposition processes in parallel, assuming that the films first decompose into stratified (layered) films before they evolve lateral structures. Such a setting is described employing a long-wave sharp-interface two-layer model. We also use a one-domain diffuse interface model to analyse the process. After describing the linear stability of stratified films in both cases we lay out some advantages and disadvantages of the two models. We conclude by mentioning some other cases of films of complex liquids, providing references for further study and discussing future challenges.

1 Introduction

Thin films of simple and complex liquids are highly relevant ingredients in a number of natural and industrial processes (Oron et al., 1997). As tear film they serve as cleaning and protection agent for eyes (Sharma and Ruckenstein, 1985); they also form a protective cover on the inside of the lungs (Borgas and Grotberg, 1988), are instrumental in bio-adhesion (Gallez and Coakley, 1996), or help as films of sweat to regulate the body temperature. In technical applications liquid films are important either because they are used during intermediate stages of an industrial production process (as in painting or coating technologies) or because they facilitate the proper working of devices (e.g., as lubricating films or in falling film evaporators). In all the mentioned applications it is important that the films remain homogeneous and do not rupture or undergo other structuring processes. For the tear film, for instance, rupture is suppressed by blinking.

However, a more recent aim consists in the usage of heterogeneous or structured films to produce hybrid, functionalised or responsive surfaces. Systems includes structured layers of bio-molecules (Mertig et al., 1997), switchable polymer brushes (Ryan et al., 2005), structured thin layers of polymer blends induced by heterogeneous substrates (Böltau et al., 1998) or electrohydrodynamic instabilities (Morariu et al., 2003), regular and irregular large-area nanoparticle structures deposited by receding contact lines (Xu et al., 2007; Pauliac-Vaujour et al., 2008). Although the emergence of actual industrial applications for structured layers produced from thin liquid films appears to be slow it is to expect that this aspect will gain increasing importance in the future.

The drive towards further miniaturisation of fluidic systems results in the fast development of micro- (Squires and Quake, 2005) and even nanofluidic devices (Mijatovic et al., 2005) as, e.g., cell asset systems employed in (medical) cell biology (Lindström and Andersson-Svahn, 2010). This implies that the understanding of the various interfacial effects will become even more important. Many of the above examples do not involve simple but rather complex liquids (Pfohl et al., 2003). This makes the understanding of the interaction of internal degrees of freedom of complex liquids with static and dynamic interfaces a practically important challenge.

The present text aims at providing some starting points for the study of dewetting of simple and some complex liquids. The individual sections give brief accounts and summarise recent results. First, Section 2 reviews the case of dewetting one-layer films of simple liquids. This system is often used as a reference case for studies of simple liquids in more complex situations or of complex liquids. Then we discuss films of binary mixtures that might undergo dewetting and decomposition processes in parallel. Experimental examples can be found in Geoghegan and Krausch (2003). Often these films first decompose into a stratified (layered film) and later evolve lateral structures. We discuss such a setting in a long-wave sharp-interface two-layer model in Section 3, and in a one-domain diffuse interface model (model-H) in Section 4. The linear stability of stratified systems is discussed in both cases. Advantages and disadvantages of the two models are discussed. Finally, we mention in the concluding Section 5 other cases of films of complex liquids, provide references for further study and discuss future challenges. Thereby, we highlight a recent derivation of a thin film model from model-H of Section 4 (Náraigh and Thiffeault, 2010) and point out that it may be written in the same form as the long-wave two-layer model from Section 3 (Thiele, 2011b).

Note, that the contribution focuses entirely on mesoscopic deterministic continuum models for thin films as derived from phenomenological transport equations. Their 'status' as continuum theory is briefly discussed in Thiele (2011c). For other approaches as e.g. (kinetic) Monte Carlo models, lattice gas methods, molecular dynamic simulations, and (dynamical) density functional theory we refer the reader to the literature (for starting points see, e.g., Rothman and Zaleski (1994); Panagiotopoulos (2000); Rabani et al. (2003); Vancea et al. (2008); De Coninck and Blake (2008); Thiele et al. (2009b); Archer et al. (2010); Léonforte et al. (2011), references therein and citing papers).

2 Dewetting of a single layer of simple liquids

2.1 General aspects

The importance of coatings for many areas has led to a large and permanent interest in the process of dewetting of a liquid film on a solid substrate. For reviews see, e.g., de Gennes (1985); Bonn et al. (2009); Craster and Matar (2009).

A typical dewetting experiment proceeds like this: A smooth solid substrate is coated by a liquid layer, for instance, by dip- or spin-coating. If the sum of the liquid-gas γ_{lg} and solid-liquid γ_{sl} interface tensions is larger than the solid-gas interface tension γ_{sg} the liquid film is (linearly or non-linearly) unstable and will eventually rupture. After some dynamical process the liquid will be collected in drops whose free surfaces have the shape of spherical caps. Each cap-like liquid-gas interface meets the solid substrate forming an equilibrium contact angle θ_e. The (macroscopic) angle is given by the Young-Laplace law $\gamma_{lg}\cos\theta_e = \gamma_{sg} - \gamma_{sl}$. Although the macroscopic picture is rather clear (de Gennes, 1985), more care is necessary when studying the pathways of film rupture and subsequent time evolution (Thiele, 2003). Then mesoscale effects have to be considered (de Gennes, 1985; Pismen, 2002). Note that many theories assume non-volatile liquids. For discussions of the volatile case see, e.g., Lyushnin et al. (2002) and Starov and Velarde (2009).

In a paradigmatic experiment Reiter (1992) deposits a ultra-thin polymer film with a thickness below 100 nm through spin-coating of a polymer solution on a solid substrate. The solvent evaporates and a smooth solid polymer film remains. When the film is brought above the glass transition temperature of the polymer, it ruptures either through heterogeneous nucleation or via an exponentially growing linear surface instability (Thiele, 2003; Seemann et al., 2005). The latter process was called spinodal dewetting by Mitlin (1993). The process results either in a labyrinth (or spinodal) structure or in individual drops or holes (Sharma and Khanna, 1998). In the latter case the holes grow resulting in a network of liquid rims. The rims may later decay via a Plateau-Rayleigh instability into rows of small drops (Reiter, 1992) that then coarsen on a much longer time scale. Sometimes the receding dewetting front is transversely unstable and leaves fingers or droplets behind (Sharma and Reiter, 1996; Reiter and Sharma, 2001). For a more detailed account and more references see Thiele (2007, 2010).

To model the dynamics of such a dewetting process in detail, the macroscopic description based on the Young-Laplace law and macroscopic hydrodynamics is not well suited. Instead, one models the process based on mesoscopic hydrodynamics, i.e., one amends the macroscopic description of the Navier-Stokes or Stokes equation to account for effective molecular forces that act between the film surface and the substrate (de Gennes, 1985; Bonn et al., 2009). These forces are related to the wettability of the substrate for the liquid (under the ambient medium – here air) and are sometimes called "surface forces" (Starov and Velarde, 2009). They enter the hydrodynamical description as a so-called disjoining or Derjaguin pressure $\Pi(h)$ (de Gennes, 1985; Israelachvili, 1992; Starov and Velarde, 2009). It was originally introduced by Derjaguin and coworkers when studying the forces between two solid bodies separated by a thin liquid film (Dzyaloshinskii et al., 1960). The disjoining pressure has a long-range part that reaches up to about 100 nm. It corresponds to apolar London–van der Waals dispersion forces (Ruckenstein and Jain, 1974). However, there are also short-range contributions acting for film thicknesses below 10nm (de Gennes, 1985; Sharma, 1993a; Pismen, 2001).

If one is interested in the static shape and stability of drops, fronts and ridges on homogeneous (or heterogeneous) substrates alone, one may obtain this information employing variational calculus. Then one only needs to know the underlying free energy functional (de Gennes, 1985; Brinkmann and Lipowsky, 2002). Note that the method is not restricted to small equilibrium contact angles. However it is not able to describe any dynamic phenomenon. For instance it can not be used to determine the dynamically most unstable linear mode for unstable states.

Being interested in the various pathways that the evolution of the film profile may take, one needs to describe the hydrodynamics of the system. One may either solve the full momentum transport (Navier-Stokes or Stokes) equations with appropriate boundary conditions at the substrate and the moving free surface. This is done, e.g., for the rupture of liquid films on heated substrates (but without accounting for wettability) in (Krishnamoorthy et al., 1995; Boos and Thess, 1999). The method is computationally expensive and does not lent itself to a systematic understanding of the system behaviour in a multi-dimensional parameter space. Alternatively, one may use an asymptotic expansion to derive a reduced model. The particular expansion often employed is the long-wave or lubrication approximation (Oron et al., 1997). It is based on the introduction of a smallness parameter $\varepsilon = l/L$ where l and L are typical length scales perpendicular and parallel to the substrate, respectively. For droplets of partially wetting liquids these would typically be the drop height and radius, i.e., the smallness parameter corresponds approximately to the equilibrium contact angle, $\varepsilon = O(\theta_e)$.

2.2 Single thin film equation

Detailed descriptions of the long-wave expansion for various situations can be found in books and review articles (Oron et al., 1997; Thiele, 2007; Craster and Matar, 2009). The present chapter builds on the derivation given in Thiele (2007).

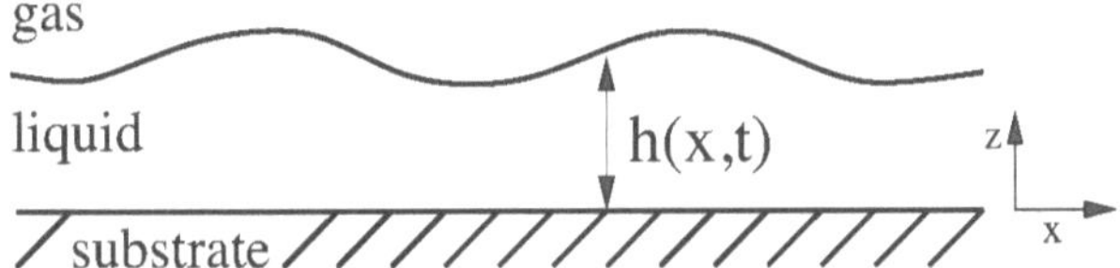

Figure 1. Sketch of the geometry of a one-layer liquid film on a solid substrate.

In general, all thin films of simple liquids are governed by the Navier-Stokes equations with adequate boundary conditions at the substrate and the free surface, i.e. the liquid-gas interface. At the liquid-gas interface the balance of the stresses is used assuming usually a passive gas. At the substrate one often applies the no-slip condition. However, also a variety of slip conditions are used like, for instance, the Navier slip condition (Oron et al., 1997; Münch et al., 2005). A kinematic boundary condition at the free surface furthermore ensures that the material boundary moves with the velocity of the liquid at the surface. Normally, the system of equations and boundary conditions is non-dimensionalised before employing the long wave approximation (Oron et al., 1997; Thiele, 2007). Depending on the details of the problem at hand and the chosen length, time, pressure and velocity scales this introduces a set of dimensionless numbers.

If a smallness parameter ε can be introduced as described above, continuity implies that not only the ratio of perpendicular and parallel length scales is small but also the ratio of the velocities perpendicular and parallel to the substrate. This allows for a simplification of the governing equations and boundary conditions. Expanding the velocity and pressure fields in ε one can solve the governing equations order by order in ε. If the dimensionless numbers are properly defined (Oron et al., 1997; Thiele, 2007), for creeping flow under weak forcing one may already at lowest order in ε obtain a thin film equation that contains all the relevant physics. If inertia matters one needs to go to higher orders and obtains, e.g., the Benney equation (Benney, 1966; Scheid et al., 2005) or integral boundary layer models (Ruyer-Quil and Manneville, 1998; Kalliadasis, 2007).

In the particular case we are interested in, we obtain a fourth order parabolic nonlinear partial differential equation describing the time evolution of the film thickness profile for a film of non-volatile liquid. We focus on the physically two-dimensional situation where the film thickness profile $h(t,x)$ depends on one spatial coordinate only (Fig. 5). To underline the physical meaning of the individual terms we write the equation in dimensional form (but in long-wave scaling):

$$\partial_t h = -\partial_x \left\{Q(h)\, \partial_x \left[\gamma \partial_{xx} h + \Pi(h)\right]\right\} \tag{1}$$

where $Q(h) = h^3/3\eta$ is the mobility function. The disjoining pressure may be written as the derivative of a local free energy $\Pi(h) = -\partial_h f(h)$. The use of a slip boundary condition at the substrate does for weak slip only change $Q(h)$, but has no effect on the equation otherwise (Oron et al., 1997). This is not the case for strong slip (Münch et al., 2005). For a discussion of dynamical thin film models including evaporation see Ajaev (2005); Rednikov and Colinet (2010); Thiele (2010) and Todorova et al. (2011).

Eq. (1) has the form of a conservation equation, i.e., the change in time of a field equals the negative divergence of a flux,

$$\partial_t h = -\partial_x \Gamma. \tag{2}$$

The flux is given as the product of a mobility and a pressure gradient, $\Gamma = Q\partial_x p$. Here, the pressure contains the laplace or curvature pressure $(\gamma \partial_{xx} h)$ and the disjoining pressure. This implies that Eq. (1) describes the evolution of a drop or film under the sole influence of capillarity and wettability. Note that for chemically inhomogeneous substrates wettability might depend on position. This can be modelled by an explicitly space-dependent $\Pi(h,x)$ (Thiele et al., 2003; Konnur et al., 2000).

It is instructive to note that Eq. (1) can be written in variational form (Mitlin, 1993)

$$\partial_t h = \partial_x \left\{Q(h)\, \partial_x \frac{\delta F[h]}{\delta h}\right\} \tag{3}$$

with $\delta/\delta h$ denoting functional variation with respect to h and the free energy functional

$$F[h] = \int \left[\frac{\gamma}{2}\,(\partial_x h)^2 + f(h)\right]\,dx. \tag{4}$$

One may say that Eq. (3) corresponds to the simplest possible equation for the evolution of a conserved order parameter field that follows gradient dynamics. The system is called "variational" (or "relaxational") and $F(h)$ represents a Lyapunov functional because it fulfils $dF/dt \leq 0$ (Langer, 1992; Mitlin, 1993). Another prominent representative of this class of systems is the Cahn-Hilliard equation describing the evolution of a concentration field for a decomposing binary mixture (Cahn, 1965; Langer, 1992). Some of the models for volatile liquids also allow for a variational formulation (Thiele, 2010).

To describe drops of a partially wetting liquid one needs a disjoining pressure that has a minimum corresponding to the thickness of a precursor layer. Examples are

$$\Pi(h) = \kappa\left[\frac{1}{h^3} - \frac{b}{h^6}\right] \tag{5}$$

and

$$\Pi(h) = \kappa\left[-\frac{b}{h^3} + e^{-h}\right]. \tag{6}$$

Eq. (5) is derived from a diffuse interface theory with non-local interactions (Pismen, 2001) whereas Eq. (6) combines a long-range apolar van der Waals interaction (that also dominates at very small scales) and a short-range polar (electrostatic or entropic) interaction (de Gennes, 1985; Sharma, 1993b,a). Another form, $\Pi(h) = 2\kappa\, e^{-h}\left(e^{-h} - 1\right)$, was derived by Pismen and Pomeau (2000) combining the long wave approximation for thin films with a diffuse interface model for the liquid-gas interface (Anderson et al., 1998). Also density variations close to the solid substrate enter their calculation in a natural way. Various other forms of disjoining pressures are used in the literature. Many results obtained employing the different forms are similar to each other if the used pressures describe a partially wetting situation and allow for a stable precursor film.

Thin film equations in the form (3) can be used to study a wide variety of phenomena in thin films on horizontal substrates. Additional effects may be incorporated into the energy functional. Examples include the influence of an electrical field for a film of dielectric liquid in a capacitor (Lin et al., 2002; Thiele and John, 2010), hydrostatic effects (Oron et al., 1997; Burgess et al., 2001), and the influence of a heated or cooled substrate (Burgess et al., 2001; Oron and Rosenau, 1992; Thiele and Knobloch, 2004). Note that the variational form of the latter problem is remarkable as there the functional governs the shape of steady liquid dro although it is out of thermodynamic equilibrium (and sustains a stationary convection cell inside the drop) (Oron and Rosenau, 1992; Thiele and Knobloch, 2004).

Beside the mentioned systems on a horizontal substrate, the long-wave approximation is used to derive dynamical equations for the evolution of film thickness and drop profiles in many other systems. This includes falling films on heated smooth (Scheid et al., 2005) or porous (Sadiq and Usha, 2008; Thiele et al., 2009a) solid substrates, films and drops on slightly inclined substrates (Oron et al., 1997; Thiele et al., 2001b), and two-layer films in the gap between two solid plates (Merkt et al., 2005; Thiele and John, 2010). All of them are modelled by single equations of conserved form (2). However, most of them do not represent gradient dynamics for a free energy functional, i.e., they can not be written in the form (3).

The long-wave approximation may also be employed to describe systems were several fields evolve in a coupled way. Examples involving liquid films are the evolution of multi-layer films (Pototsky et al., 2006), surfactant-covered films (Borgas and Grotberg, 1988; Jensen and Grotberg, 1992; Matar and Craster, 2009), films with an reactive solute or surfactant(s) (de Gennes, 1998; John et al., 2005; Pereira et al., 2007), and films of laterally decomposing mixtures Náraigh and Thiffeault (2010); Thiele (2011b). We will come back to such systems in sections 3 and 5. However, in the next section we will give a selection of basic results for single-layer systems that follow a gradient dynamics.

2.3 Basic analysis

Usually, one bases the analysis of the thin film equation (1) on (i) an analytical determination of flat film solutions and their linear and nonlinear stability properties; (ii) on a numerical analysis of steady solutions, their bifurcations and linear stability; and (iii) on studies of dynamical pathways employing time simulation techniques. For an overview and recent developments see (Thiele, 2007; Beltrame and Thiele, 2010). As this has been presented by many authors, here we only sketch main results following closely the section on dewetting of the recent review by Thiele (2010).

We employ the particular choice

$$f(h) = -\kappa\left[\frac{1}{2h^2} - \frac{b^3}{5h^5}\right] \tag{7}$$

(Pismen, 2001) for the energy term corresponding to wettability. Note that the qualitative behaviour mainly depends on the number and relative depth/height of its extrema. The present f has only one minimum

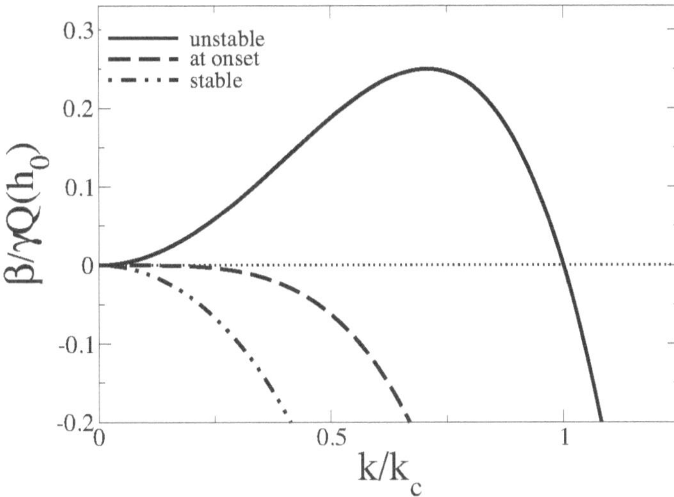

Figure 2. Dispersion relations for the surface instability of a flat liquid film w.r.t. surface modulations resulting in dewetting and the evolution of patterns of droplets. Shown is a stable, an unstable and the neutrally stable case. Figure adapted from Thiele (2010).

at $h_{\rm precursor} = b$ and in consequence may describe a precursor film of thickness b in co-existence with large drops of equilibrium contact angle $\theta_{\rm eq} = (-3\kappa/5b\gamma)^{1/2}$ (corresponding to the slope at the inflection point of the profile). With a second minimum at larger finite thickness one may also describe pancake-like drops (de Gennes, 1985; Thiele et al., 2002). However, then the determination of precursor film thickness, pancake thickness and equilibrium contact angle amounts to a Maxwell construction Thiele et al. (2002).

Inspecting Eq. (1) one notices that any flat film corresponds to a steady state solution. To determine their stability in time we linearise about the flat film of thickness h_0 employing harmonic modes, i.e., $h(x,t) = h_0 + \epsilon \exp(\beta t + ikx)$ where $\epsilon \ll 1$, β is the growth rate of the mode of wave number k. Linearising Eq. (1) gives the dispersion relation

$$\beta(k) = -Q(h_0)\,\gamma\,k^2\,(k^2 - k_c^2), \tag{8}$$

where the critical wave number is given by $k_c = \sqrt{-\partial_{hh}f|_{h=h_0}/\gamma}$. For $\partial_{hh}f|_{h=h_0} < 0$ the film is linearly unstable for $0 < k < k_c$. The fastest growing mode has $k_{\rm max} = k_c/\sqrt{2}$ and $\beta_{\rm max} = Q(h_0)\gamma k_c^4/4$. The onset of the instability occurs at $\partial_{hh}f|_{h=h_0} = 0$ with $k_c^{\rm onset} = 0$, i.e., it is a long-

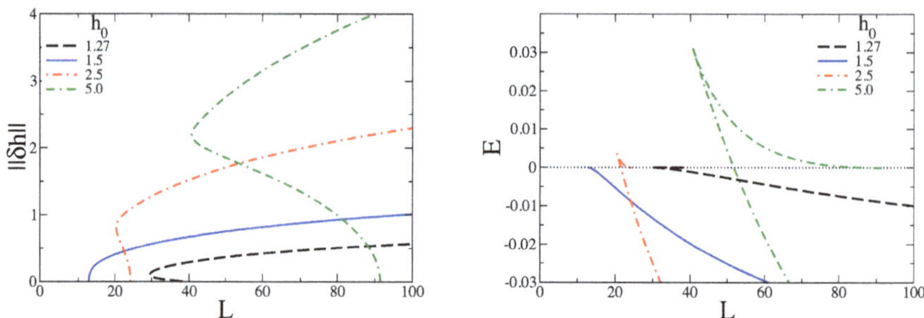

Figure 3. We characterise families of steady one-dimensional drop solutions by (left) their L_2-norm $||\delta h||$ and (right) the energy F (Eq. (4)) per length. The legends give the corresponding mean film thicknesses. Figure adapted from Thiele (2010).

wave instability. For the present disjoining pressure (7) one finds a critical film thickness $h_c = (2)^{1/3}b$. The film is linearly unstable for $h > h_c$. For examples of dispersion curves (8) see Fig. 2. Note that mass conservation implies $\beta(k=0) = 0$.

Steady profiles are obtained by solving Eq. (1) with $\partial_t h = 0$. Taking domain size L (or period) as control parameter and fixing the mean film thickness h_0, for a linearly unstable film a one parameter family of profiles bifurcates from the flat film at $L_c = 2\pi/k_c$ (Fig. 3). Some corresponding drop profiles are given in Fig. 4. We characterise solutions by their norm $||\delta h|| = [(1/L)\int_0^L (h(x) - h_0)^2 dx]^{1/2}$ and energy per length $E = (1/L)\int_0^L F[h]dx$.

Note that branches of steady solutions bifurcate super- or subcritically from the flat film solution. In the latter case the subcritical part of the branch consists of unstable nucleation solutions that resemble holes. They are important for film rupture if the system is noisy or 'dirty': In the evolution of defect-ridden films they offer a fast track to film rupture. This leads to the distinction of nucleation-dominated and instability-dominated behaviour of linearly unstable thin films (Thiele et al., 2001a; Thiele, 2003). For the analysis of a three-dimensional system see Beltrame and Thiele (2010).

The respective branches that bifurcate at L_c (Fig. 3) are the first of a respective infinite number of primary solution branches. They bifurcate at domain sizes $L_{cn} = 2\pi n/k_c$, $n = 1, 2, \ldots$. The branch bifurcating at L_{cn} corresponds to the $n = 1$ branch 'stretched in L' by a factor n. The individual thickness profiles on the n branch consist of n identical drops.

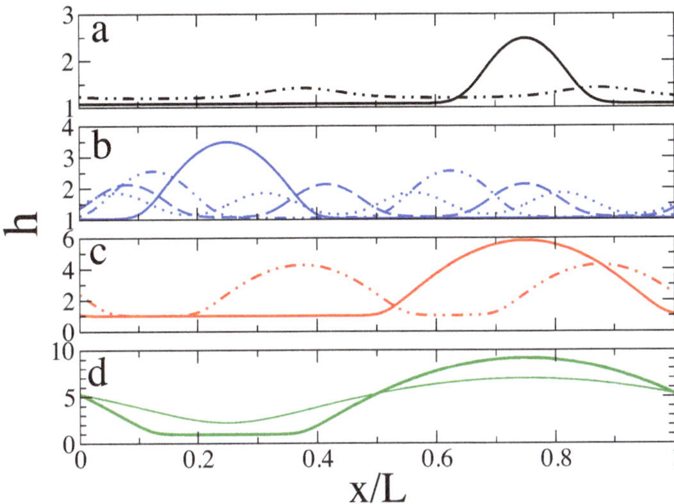

Figure 4. Examples of drop profiles for several branches presented in Fig. 3. Mean thicknesses $\bar{h}$ are (a) 1.27, (b) 1.5, (c) 2.5, and (d) 5.0. The system size is $L = 60$ in all cases. Given are profiles from the following branches: $n = 1$ (solid lines), $n = 2$ (dot-dashed lines), $n = 3$ (dashed line), and $n = 4$ (dotted line). The thin line in (d) corresponds to a nucleation solution, i.e., to a critical hole. Note that for any given L only the $n = 1$ profiles are stable w.r.t. coarsening. Figure adapted from Thiele (2010).

Note that in the present problem, the different branches are decoupled. However, the situation changes for other energy functionals (Thiele, 2010) or when the reflection or translational symmetry of the system is broken (by including a lateral driving force (Golovin et al., 2001; Thiele et al., 2001b; Thiele and Knobloch, 2004), substrate heterogeneity (Thiele et al., 2003), or lateral boundary conditions (Thiele et al., 2007)).

Here we do not discuss the nonlinear stability of flat films and the related steady state solutions on branches not connected to the flat film solutions (but see e.g. Thiele (2003); Thiele and Knobloch (2004)). For studies of the time evolution of dewetting films in two- and three-dimensional settings we refer to Sharma and Khanna (1998); Bestehorn and Neuffer (2001); Becker et al. (2003); Verma et al. (2005) and Beltrame and Thiele (2010). Next, we turn our attention to thin films of binary mixtures. Such films might be unstable with respect to dewetting of the film *and* to decomposition of the mixture. If the mixture decomposes it might first decompose perpendicularly to the substrate forming a layered structure and on a larger time scale structure as well laterally. The latter process may be studied directly, employing a long-wave model for two-layer films as considered in the next section.

3 Two-layer films

The previous section has given a brief overview of the thin film description of one-layer films, in particular, in relation to dewetting. The situation becomes more intricate for two-layer films as sketched in Fig. 3. A number

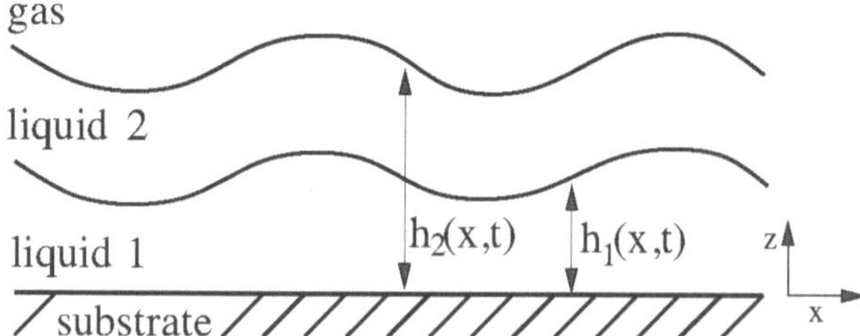

Figure 5. Sketch of the geometry of two-layer film on a solid substrate.

of experiments studies different aspects of the dewetting of two-layer films where the two involved liquids are immiscible and the two layer structure is imposed from the beginning (David et al., 1998; Pan et al., 1997; Sferrazza et al., 1998; Lin et al., 2002). Investigated effects are interface instabilities or the growth of holes. These experiments can be directly described by a sharp-interface long-wave two-layer model (Pototsky et al., 2005; Bandyopadhyay et al., 2005).

Other experiments investigate the evolution of thin films of mixtures on a solid substrate (Geoghegan and Krausch, 2003). The films might decompose in the direction perpendicular to the substrate forming a two-layer film. If they are thick as compared to the length scale of decomposition they might form a normal spinodal pattern or multi-layer patterns (via so-called "spinodal decomposition waves" or "surface modes") (Plapp and Gouyet, 1997). The underlying processes can not be described employing a long-wave theory. Instead one may use a diffuse-interface description as model-H (see Section 4). If the film thickness is about or below the length scale of decomposition both – the sharp-interface long-wave two-layer model and the diffuse-interface short-wave one-domain model – are able to describe certain aspects of the system. The present Section focuses on the former and Section 4 introduces the latter.

To introduce the long-wave two-layer model we limit our attention to a two-dimensional physical situation as sketched in Fig. 5. The following presentation of the evolution equations and linear stability analysis follows closely Pototsky et al. (2005). For results on the non-linear evolution, in particular, on different pathways of coarsening we refer to Pototsky et al. (2005) and Pototsky et al. (2006) for the two- and three-dimensional case, respectively.

As for the one-layer film a long-wave approximation is applied to the momentum transport equations, i.e., here to the Navier-Stokes equations in liquid 1 and 2 and to the boundary conditions at the substrate, the liquid-liquid interface and the free surface. As a result one obtains coupled evolution equations for the film thicknesses $h_1(x,t)$ and $h_2(x,t)$. They can be written in variational form (as gradient dynamics)

$$\begin{aligned}\frac{\partial h_1}{\partial t} &= \partial_x\left(Q_{11}\partial_x\frac{\delta F}{\delta h_1} + Q_{12}\partial_x\frac{\delta F}{\delta h_2}\right)\\ \frac{\partial h_2}{\partial t} &= \partial_x\left(Q_{21}\partial_x\frac{\delta F}{\delta h_1} + Q_{22}\partial_x\frac{\delta F}{\delta h_2}\right),\end{aligned} \tag{9}$$

i.e., in terms of the functional derivatives $\delta F/\delta h_i$ (with $i = 1,2$) of the energy functional

$$F = \int\left(\frac{\gamma_{12}}{2}(\partial_x h_1)^2 + \frac{\gamma_2}{2}(\partial_x h_2)^2 + f(h_1,h_2)\right)dx. \tag{10}$$

One may also say that the system (9) represents the general form of coupled evolution equations for two conserved order parameter fields in a relaxational situation. For a discussion of related models see Pototsky et al. (2005).

In Eq. (10) the parameters γ_{12} and γ_2 are the liquid-liquid and liquid-gas interface tensions, respectively. The local free energy per unit area $f(h_1,h_2)$ reflects the van der Waals interactions between the four materials: substrate, liquid 1, liquid 2 and ambient gas. In a partially wetting situation they destabilise flat thin films and lead to dewetting. Inclusion of other interactions, like for instance, stabilising short-range interaction, is straightforward (Pototsky et al., 2005) and necessary if one plans to study the long-time evolution with Eqs. (9).

The mobility functions $Q_{ik}(h_1,h_2) \geq 0$ are the components of the symmetric mobility matrix

$$\mathbf{Q} = \frac{1}{\eta_1}\begin{pmatrix} \frac{h_1^3}{3} & \frac{h_1^2}{2}(h_2 - \frac{h_1}{3}) \\ \frac{h_1^2}{2}(h_2 - \frac{h_1}{3}) & \frac{(h_2-h_1)^3}{3}(\frac{\eta_1}{\eta_2} - 1) + \frac{h_2^3}{3} \end{pmatrix}. \tag{11}$$

The symmetry $Q_{12} = Q_{21}$ corresponds to an Onsager reciprocal relation on the level of the thin film equations (9) where the gradients $\partial_x(\delta F/\delta h_i)$ represent the 'thermodynamic forces'.

The functional F corresponds to a Lyapunov functional, i.e., it decreases monotonously in time: The total time derivative is

$$dF/dt = \int \left(\frac{\delta F}{\delta h_1} \partial_t h_1 + \frac{\delta F}{\delta h_2} \partial_t h_2 \right) dx. \tag{12}$$

Employing Eq. (9) and partial integration one obtains

$$\frac{dF}{dt} = - \int \sum_{i,k} Q_{ik} \left(\partial_x \frac{\delta F}{\delta h_i} \right) \left(\partial_x \frac{\delta F}{\delta h_k} \right) dx. \tag{13}$$

Because $\det \mathbf{Q} > 0$ and $Q_{11}, Q_{22} > 0$, the quadratic form in Eq. (13) is positive definite implying $dF/dt < 0$. The steady state solutions of Eqs. (9) can be obtained by a minimisation of F.

Two-layer films can (as one-layer films) rupture via a surface instability or via nucleation. The latter effect has to our knowledge not yet been investigated. The surface instability can be studied systematically with a linear stability analysis of two-layer flat films. In this way, one is able to answer intricate question for the first stage of dewetting as, e.g., "*Which* interface will dominate the instability?" and "*Where* does the film rupture?"

The stability of flat two-layer films is studied employing the ansatz $h_i(x) = d_i + \epsilon\chi_i \exp(\beta t) \exp(kx)$ for $i = 1, 2$ where d_i, k, β and $\epsilon\chi = \epsilon(\chi, 1)$ are the flat film heights, wave number, growth rate and amplitudes of the disturbance, respectively. Linearising Eqs. (9) in $\epsilon \ll 1$ results in the eigenvalue problem $(\mathbf{J} - \beta\mathbf{I})\chi = 0$ with the non-symmetric Jacobi matrix $\mathbf{J} = -k^2\mathbf{Q} \cdot \mathbf{E}$. The energy matrix is

$$\mathbf{E} = \begin{pmatrix} \frac{\partial^2 f}{\partial h_1^2} + k^2 & \frac{\partial^2 f}{\partial h_1 \partial h_2} \\ \frac{\partial^2 f}{\partial h_1 \partial h_2} & \frac{\partial^2 f}{\partial h_2^2} + \sigma k^2 \end{pmatrix}, \tag{14}$$

and $\mathbf{Q}$ is the scaled mobility matrix (Pototsky et al., 2005).

Solving for the eigenvalues yields the dispersion relation

$$\beta(k) = \frac{\mathrm{Tr}}{2} \pm \sqrt{\frac{\mathrm{Tr}^2}{4} - \mathrm{Det}}, \tag{15}$$

where Tr$= -k^2[2Q_{12}E_{12} + Q_{11}E_{11} + Q_{22}E_{22}]$ and Det$= k^4 \det \mathbf{Q} \det \mathbf{E}$ are the trace and the determinant of $\mathbf{J}$, respectively. Since $\det \mathbf{Q} \neq 0$ the eigenvalue problem can be written as the generalised eigenvalue problem $(k^2\mathbf{E} + \beta\mathbf{Q}^{-1})\chi = 0$. Because $\mathbf{E}$ and $\mathbf{Q}^{-1}$ are both symmetric and $\mathbf{Q}^{-1}$ is positive definite all eigenvalues β are real as expected for a variational problem. A further implication is that the instability threshold (onset) is determined by the eigenvalues of $\mathbf{E}$ alone and does not depend on the mobilities. The onset of the instability is always at $k = 0$, i.e. the system is linearly stable for

$$\det \mathbf{E} > 0 \quad \text{and} \quad E_{11} > 0 \quad \text{at} \quad k = 0, \tag{16}$$

independently of the wavelength of the disturbance.

Performing the linear analysis for different physically realistic parameter sets (Pototsky et al., 2005) one finds two different types of unstable modes. These are zigzag ($\chi > 0$ at k_{max}, Fig. 6(a)) and varicose ($\chi < 0$ at k_{max}, Fig. 6(b)) modes, i.e., the two interfaces move in phase and in anti-phase, respectively. The linear modes also strongly influence the non-linear evolution towards rupture. Without incorporation of stabilising short-range interactions the zigzag and varicose mode leads (for parameters chosen as in Pototsky et al. (2005)) to rupture at the liquid-liquid interface and at the substrate, respectively. Both modes are normally asymmetric ($\chi \neq 1$), i.e., the deflection amplitudes of the two interfaces are normally different. Note that although the mobilities have no influence on the stability threshold, they determine the mode type and the length and time scales of the evolution.

Beside the two well known modes one may find a mixed mode, i.e., a dispersion relations with two maxi ma. If they are of about equal height as in Fig. 6(c) two modes of different wavelength may evolve at the two interfaces, respectively. For this to occur $\chi \ll 1$ and $\chi \gg 1$ at the two k_{max}, respectively. Such mixed modes may explain the two length scales visible in Fig. 29(e) of Geoghegan and Krausch (2003).

The model presented here is restricted because it only includes long-range van der Waals interaction that act destabilising. Therefore every evolution of a surface structure ends with the rupture of one of the two layers. To be able to study different pathways of coarsening that may occur in the long-time evolution one has to include stabilising short-range interactions that allow for thin precursor films (Pototsky et al., 2005, 2006).

4 Decomposing free surface films

We have seen in the preceding sections that long-wave sharp-interface models are well developed for one-layer and two-layer films of simple liquids. Although the two-layer model may describe aspects of the evolution of films

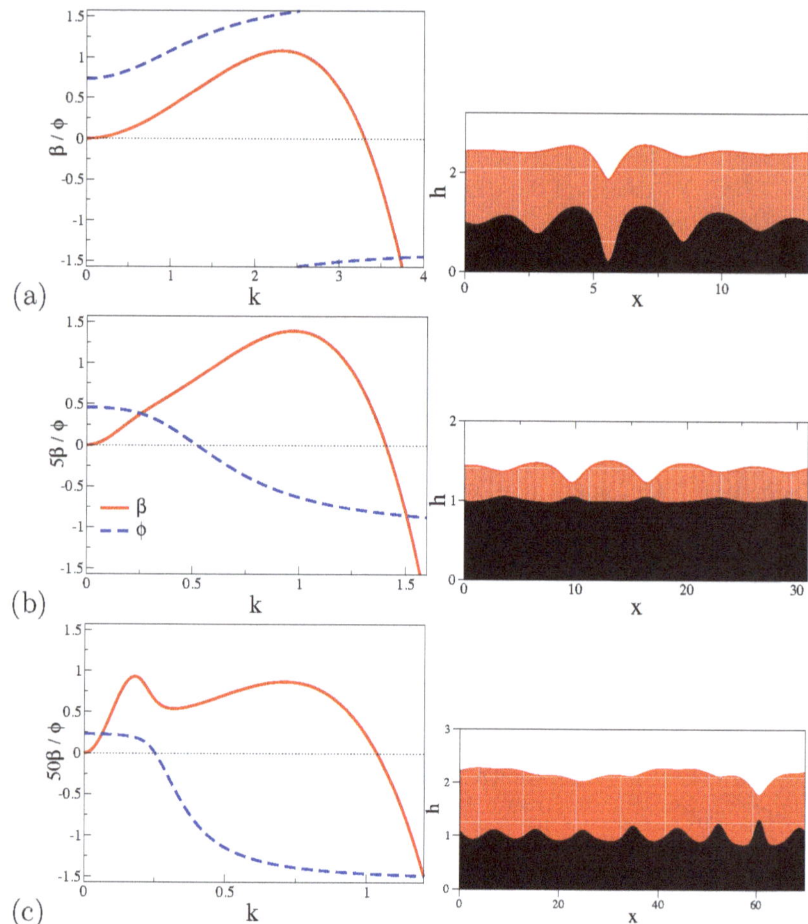

Figure 6. Linear and nonlinear results are shown for of a two-layer film for a Si/PMMA/PS/air system. The left panels show the growth rate β (solid lines) and the mode type $\phi = \chi/(1+\chi^2)$ (dashed lines) of the leading ignored in dependence of the wave number k. Negative ϕ [positive ϕ] corresponds to varicose [zigzag] modes. For convenience we plot in (a) 20β and in (b) 10β. The right panels give corresponding snapshots from time evolutions. (a) At $d = d_2/d_1 = 2.4$, $\mu = 1.0$ a zigzag mode evolves and rupture of the lower layer occurs at the substrate. (b) At $d = 1.4$, $\mu = 1.0$ a varicose mode evolves leading to rupture of the upper layer at the liquid-liquid interface. (c) At $d = 1.4$ as in (b) but for $\mu = 0.1$ a mixed mode evolves and rupture occurs as in (b). The domain lengths are 5 times the corresponding fastest unstable wave length and $\sigma = 1$ (for more details see Pototsky et al. (2005)).

of decomposing mixtures, such models are only valid if several conditions are fulfilled. These include that (i) an initial vertical stratification occurs much faster than any subsequent lateral structuring, (ii) the resulting liquid-liquid interface is 'sufficiently sharp' and (iii) no enrichment or depletion zones develop at the free surface of the film or at the substrate.

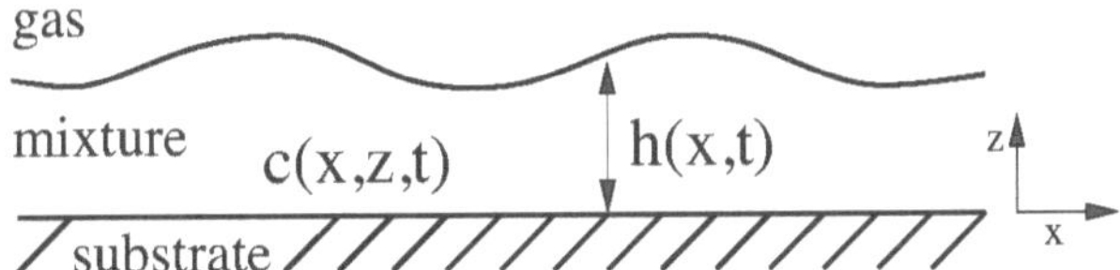

Figure 7. Sketch of the geometry of free surface film of a liquid mixture on a solid substrate.

These conditions may hold in part of the experiments reviewed, e.g. in Geoghegan and Krausch (2003). They do clearly not hold in others (Oron et al., 2004) where decomposition and surface structuring interact. To model such systems one needs a model that is able to describe decomposition of the mixture and surface structuring of the film (see Fig. 7). This is not possible with either thin film theories for simple liquids as discussed above or models for decomposition in a gap of fixed geometry (Fischer et al., 1998).

A candidate for a suitable model is based on model-H, i.e., on coupled transport equations for momentum and a concentration field (Anderson et al., 1998). Applied to a bulk system, this diffuse interface model is able to describe decomposition processes in viscose liquids (Vladimirova et al., 1999). However, to model a film of a mixture with a free surface the bulk equations have to be supplemented by boundary conditions at the sharp liquid-gas interface (Thiele et al., 2007). With this amendment one is able to describe the evolution of diffuse interfaces inside the film and their influence on the sharp liquid-gas interface.

In the following we will first briefly present the bulk equations and boundary conditions. Then we discuss steady stratified states and their lateral stability. We end with a discussion of the influence of hydrodynamic motion. The presentation follows closely parts of Thiele et al. (2007) and Madruga and Thiele (2009). We also refer to Frastia et al. (2011b) and Bribesh et al. (2011).

4.1 Bulk equation and boundary conditions

Model-H (Hohenberg and Halperin, 1977; Anderson et al., 1998) couples the description of decomposition (Cahn-Hilliard equation, Cahn (1965)) and of momentum transport (Navier-Stokes equations, Batchelor (2000)). To this aim one introduces a convective Cahn-Hilliard equation

$$\partial_t c + \mathbf{v}\cdot\nabla c = -\nabla\cdot\{M\nabla\left[\sigma_c\Delta c - \partial_c f(c)\right]\}. \qquad (17)$$

and couples it to amended Navier-Stokes equations

$$\rho\frac{\partial \mathbf{v}}{\partial t} + \rho\mathbf{v}\cdot\nabla\mathbf{v} = -\nabla\cdot\{\sigma_c(\nabla c)(\nabla c) + p_{\text{eff}}\,\mathbf{I}\} + \eta\Delta\mathbf{v}. \qquad (18)$$

Here, the model is written in terms of the concentration difference of the two components c and the velocity $\mathbf{v}$. The effective pressure p_{eff} contains all diagonal parts of the stress tensor. The parameters M, η, ρ and σ_c are a mobility constant, dynamic viscosity, density and a stiffness related to the interfacial tension of the diffuse liquid-liquid interface, respectively. $f(c)$ is the bulk energy density responsible for decomposition. Normally, a double well potential is used.

Note that in the literature the bulk model-H is presented in various forms. In particular, the momentum equation is written in different ways. Most differences arise from different (often implicit) definitions of the pressure p_{eff}. For more details, a discussion of the non-dimensionalisation, and 'translations' between some of the different formulations see Thiele et al. (2007).

Because of its importance for the boundary conditions we also give the stress tensor

$$\boldsymbol{\tau} = -p_{\text{eff}}\mathbf{I} - \sigma_c(\nabla c)(\nabla c) + \eta\left(\nabla\mathbf{v} + (\nabla\mathbf{v})^T\right). \qquad (19)$$

Compared to classical hydrodynamics it contains an additional concentration-gradient-dependent contribution (Korteweg stresses (Joseph, 1990)).

To describe the coupled time evolution of the film thickness and concentration profiles one has to supplement the coupled transport equations (17) and (18) with appropriate boundary conditions at the free surface and at the solid smooth substrate.

At both boundaries one prescribes zero diffusive flow through the boundary and a potential energetic preference of the boundary for one component of the mixture. This gives the needed four boundary conditions that accompany Eq. (17). Here, we assume that the solid substrate is neutral w.r.t. the components of the mixture. However, the free surface might prefer one of the components, i.e., the excess surface energy $f^s = \gamma_0 + a^s c$ depends on concentration. Here, we chose a linear dependence. For $a^s > 0$ negative c are preferred, i.e., for $c = c_1 - c_2$, component 2 is enriched at the free surface.

The conditions for the velocity field are no-slip and no-penetration at the solid substrate and the force equilibrium at the free surface

$$\boldsymbol{\tau} \cdot \mathbf{n} = -\gamma(c)\,\mathbf{n}\nabla \cdot \mathbf{n} + \mathbf{t}(\mathbf{t} \cdot \nabla\gamma(c)) \tag{20}$$

where we assume that the ambient air does not transmit any force. The expression $\nabla \cdot \mathbf{n}$ corresponds to the curvature of the free surface, i.e., the first r.h.s. term corresponds to the laplace pressure. The second term represents a solutal Marangoni force that acts tangential to the interface. In the present two-dimensional formulation, $\mathbf{n}$ and $\mathbf{t}$ represent the normal and tangent unit vector of the free surface profile, respectively. Note that we identify the surface tension $\gamma(c)$ and excess surface energy $f^s(c)$ (Thiele et al., 2007).

To describe the motion of the free surface one furthermore imposes the kinematic condition

$$\partial_t h = w - u\partial_x h \tag{21}$$

where we used $\mathbf{v} = (u, w)$. It ensures that the surface follows the flow field. Note that the final condition implies that an evolving film profile is by definition related to a hydrodynamic flow of the mixture.

After non-dimensionalisation (Thiele et al., 2007) the ratio $\sigma_c/\eta ME$ turns out to be the most important bulk parameter for creeping flow. It corresponds to the ratio of the typical velocity $U' = \sigma/\eta$ of the viscose flow driven by the internal 'diffuse interface tension' $\sigma = \sigma_c/l = \sqrt{\sigma_c E}$ and the typical velocity of diffusive processes $U = M\,E/l = D/l$. This implies that it can as well be seen as a Peclet number $U'l/D$ or as an inverse Capillary number $\sigma/U\eta$. Here, $l = \sqrt{\sigma_c/E}$ is the width of the diffuse interface, E is typical energy density resulting from the quartic polynomial of the bulk energy, and D is a diffusion constant. The non-dimensional boundary conditions contain a surface tension number $\mathrm{S} = \gamma_0/\sigma$ and a Marangoni number $\mathrm{Ma} = a^s/\sigma$. Note, that here we have not included a dependence of the dimensionless numbers on the binodal concentration as it does not change the overall set-up (but see Thiele et al. (2007)).

We are now equipped with a complete model to investigate a wide variety of systems involving decomposing mixtures with free surfaces. As an example we next sketch for the case $\sigma_c/\eta ME = 1$, $\mathrm{S} = 1$, $\bar{c} = 0$ and $\mathrm{Ma} \geq 0$ the analysis of base states and their linear stability done in the literature.

4.2 Stratified layers and their linear stability

The performed analysis first studies quiescent, vertically homogeneous and vertically stratified base state solutions (Thiele et al., 2007) and their linear stability with respect to lateral perturbations (Madruga and Thiele, 2009). The general solution of this problem may be written in the form

$\mathbf{v}(x,z,t) = \mathbf{v_0} + \varepsilon\tilde{\mathbf{v}}_1(x,z,t)$, $p_{\text{eff}}(x,z,t) = p_0(z) + \varepsilon\tilde{p}_1(x,z,t)$, $c(x,z,t) = c_0(z) + \varepsilon\tilde{c}_1(x,z,t)$, and $h(x,t) = h_0 + \varepsilon\tilde{h}_1(x,t)$. The fields $\varepsilon\tilde{\mathbf{v}}_1$, $\varepsilon\tilde{p}_1$, $\varepsilon\tilde{c}_1$, and $\varepsilon\tilde{h}_1$ denote the infinitesimal perturbations of velocity, pressure, concentration, and thickness fields, respectively. The small parameter ε is used to linearise the time-dependent equations about the quiescent base states. The studied base states correspond to a flat layer (h_0 =constant) of a quiescent fluid mixture ($\mathbf{v}_0 = 0$) where the vertical concentration profile $c_0(z)$ is a solution of the steady one-dimensional Cahn-Hilliard equation with appropriate boundary conditions. Further $p_0 = -(\partial_z c_0)^2$. In Thiele et al. (2007) these states are extensively studied in dependence of film thickness and different modi and strength of energetic bias at the two interfaces. Without bias one finds homogeneous base states where the concentration c_0 is constant across the sample *and* above a critical film thickness (of $h_c = \pi$ for the used parameter values) layered (but laterally homogeneous) films. With an energetic bias normally no homogeneous base states exist as the bias always causes an (additional) local depletion or enrichment at the corresponding interface. However, one might still distinguish weakly and strongly layered films.

Fig. 8 presents an example of a characteristics of the resulting solution branches for the case of an asymmetric bias, i.e., only the energy of the free surface depends on Ma. The norm is given in dependence of the Marangoni number (e.g., the strength of the energetic bias at the free surface). For neutral surfaces (Ma= 0) above $h_c = \pi$ two types of steady states exist: homogeneous film ($||\delta c|| = 0$) and stratified film ($||\delta c|| > 0$). For Ma> 0 the homogeneous film becomes a film with a depletion layer with $c < 0$ near the free surface. Note that at Ma= 0 there exist two stratified film solutions of identical norm: the two layered states with liquid 1 on top of liquid 2 and vice versa are related by the symmetry $z \to h - z$). As the two states react differently to the bias, the symmetry is broken by the surface bias, i.e., two curves emerge (Fig. 8). One of them (the one of lower norm and higher energy) annihilates at a finite Ma in a saddle-node bifurcation with the depletion layer solution. Physically, the most interesting solution is for each respective h the one of highest norm in Fig. 8 as it is the state of lowest energy. It continues towards large Ma. For $h < h_c$ only one solution exists for all Ma. It corresponds to a weakly layered film, i.e., the film with a depletion layer. For more details as e.g. plots of concentration profiles see Thiele et al. (2007).

We have based our analysis on the assumption that a film of a mixture may first decompose into a layered structure (our steady states c_0) before on a much larger time scale lateral structures evolve. Following this line of argument we next study the lateral stability of the base states

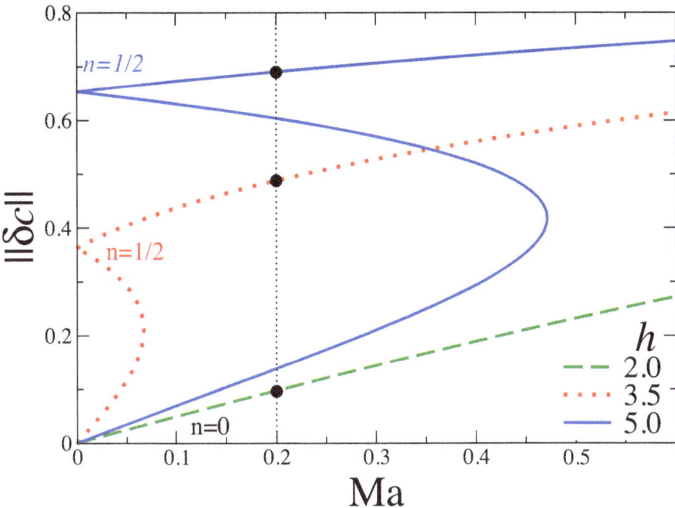

Figure 8. Branches of steady stratified flat-surface films of a liquid mixture are characterised by the L_2-norm of the concentration fields $c_0(z)$. Several layered states might exist in dependence of the film thickness and the energetic bias at the free surface (a^s) measured here by the Marangoni number Ma$= a^s/\sigma$. The mean concentration is $\bar{c} = 0$, and the film thicknesses are as given in the legend. Figure adapted from Thiele et al. (2007).

using the above ansatz: We decompose the perturbations into a sum of normal modes $(\tilde{\mathbf{v}}_1, \tilde{p}_1, \tilde{c}_1, \tilde{h}_1) = (\mathbf{v}_1(z), p_1(z), c_1(z), h_1)\exp(\beta t + ikx)$, where β is the growth rate and k the lateral wave number. Using this ansatz in Eqs. (17) to (18) one obtains linearised convective Cahn-Hilliard and momentum equations accompanied by linearised boundary conditions for the perturbations (Madruga and Thiele, 2009).

For homogeneous films at Ma$= 0$ the perturbations of the composition and velocity fields decouple and the linear stability is equivalent to the one governed by a purely diffusive Cahn-Hilliard description of a decomposing mixture in a confined slab-type geometry. It is found that the film is laterally unstable for all thicknesses (see also Fischer et al. (1998)). For the layered solutions the linear system of equations and boundary conditions does not decouple resulting in a dependence of the linear stability on hydrodynamic aspects.

As above we distinguish the cases of neutral and energetically biased free surface. In the neutral case the vertical gradient of the base state concentration is zero at both surfaces and increases into the bulk. Therefore any flow is driven from within the bulk film, i.e., by the diffuse interface between the two components. We call this type of forcing “bulk Korteweg driving”. However, with the energetic bias also the gradients in the depletion layer at the free surface can destabilise the film resulting in a “surface Korteweg (or Marangoni) driving” (Madruga and Thiele, 2009). In cases where both mechanisms are present it depends mainly on film thickness and surface bias which one is dominant.

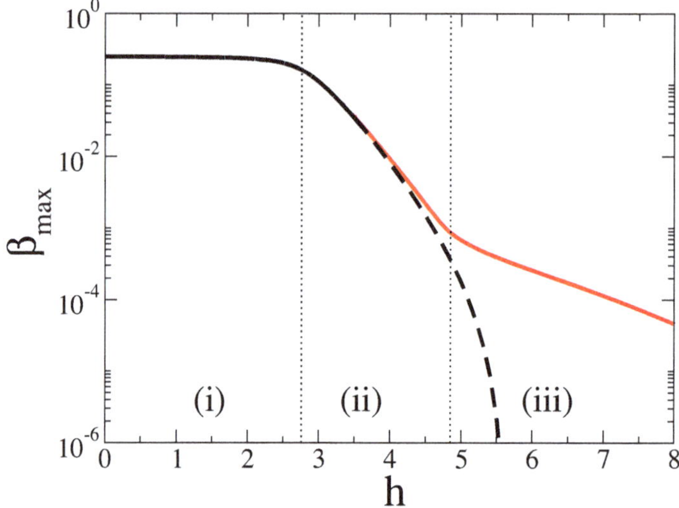

Figure 9. The maximal growth rate of the lateral instability modes is given in dependence of the mean film thickness h for a moderate energetic surface bias of Ma= 0.2. The dashed line is for purely *diffusive transport* whereas the solid line is for coupled *diffusive and convective transport*. Regions (i) to (iii) are explained in the main text. The location of the solutions for selected h values in Fig. 8 is indicated there by filled circles. Figure adapted from Madruga and Thiele (2009).

As an example, Fig. 9 analyses the growth rates of the most dangerous linear instability mode of the energetically favourable layered base state. The rate is given in dependence of mean film thickness h in the case of a moderate energetic surface bias (Ma= 0.2). The loci of relevant solutions in Fig. 8 are indicated there by the filled circles. It is instructive to compare the results obtained with the linearised full model-H to results calculated for an imposed $\mathbf{v}_1 = 0$, i.e., only considering diffusive transport of material. In the latter case, by definition no convective transport occurs and the film has to remain flat. Inspecting Fig. 9 one may distinguish three regions: (i) For small $h \lesssim 3$ the cases with and without convective transport are practically identical, and the maximal growth rate is about constant. The base states are weakly layered films. Its bulk is at $c = 0$ resulting in a nearly purely lateral instability with characteristics similar to a one-dimensional bulk mode (Madruga and Thiele, 2009). Above $h \approx 3$ in region (ii) the base state changes its character to a strongly layered film. The growth rate decreases exponentially with film thickness as a result of the smaller importance of the driving interfacial layers as compared to the bulk regions of the two layers. However, one notices that for purely diffusive driving the film actually stabilises at about $h = 5.5$ whereas the full model with convection results in a different behaviour. Already in the upper part of region (ii) the growth rate with convection becomes larger than the one without. However, in region (iii) above $h \approx 5$ the behaviour changes qualitatively. Convection keeps the film unstable as it allows for surface deformations to occur. Growth rates still decrease exponentially with increasing film thickness but at a slower pace than before.

We emphasise that the rationale behind our finding is not only that convective motion adds a second transport process beside diffusion to the dynamics. It also allows the film to change its surface profile, i.e., its geometry. It is then able to realise a different class of solutions, namely, films with laterally modulated concentration *and* thickness profiles. This qualitative change is as well reflected in a change of the character of the perturbation velocity profile. In region (ii) it is non-monotonous indicating driving from the diffuse interface inside the film whereas in region (iii) driving comes from the free surface. This corresponds to the transition from bulk Korteweg driving to surface Korteweg (or Marangoni) driving. For more details of the linear stability results see Madruga and Thiele (2009).

5 Conclusion and outlook

We hope the present text has provided some starting points for further studies of dewetting simple and complex liquids. We have first reviewed the case of dewetting of one-layer films of simple liquids as this case is often used as a reference. Here, we have only considered homogeneous substrates and static film-substrate interactions. We have further excluded thermal effects and lateral driving forces. However, many of these can efficiently be included in the thin film description. This is described in the review by Thiele (2007). Some more recent references where such extensions are developed or employed concern ratchet-driven fluid transport in bounded two-layer films (John et al., 2008); the dewetting dynamics of a thin liquid film sandwiched between a solid substrate and a bulk of a non-miscible fluid (Verma and Sharma, 2007); drops driven by substrate vibrations (John and Thiele, 2010; Thiele and John, 2010), film flow on a porous substrate (Thiele et al., 2009a); and the dynamics of drops depinning from substrate heterogeneities (Beltrame et al., 2009, 2011) and its relation to drops on a rotating cylinder (Thiele, 2011a). Extensions towards volatile liquids are discussed e.g. by Ajaev (2005); Dunn et al. (2009) and Todorova et al. (2011). Recent reviews that cover part of these aspects are Craster and Matar (2009); Thiele (2010) and Bonn et al. (2009).

Next, we have discussed the case of a film of a liquid mixture that might undergo dewetting and decomposition. Assuming that the mixture decomposes during the early stage of its evolution perpendicularly to the substrate, we have presented two approaches: A two-layer sharp-interface theory in the form of coupled evolution equations for the layer thickness profiles (derived employing long-wave approximation, see Section 3) and a diffuse-interface short-wave one-domain model based on model-H with boundary conditions at the sharp liquid-solid and liquid-gas interfaces (Section 4). The two models describe aspects of the decomposition and dewetting process in different limiting cases and can not be directly 'translated' into each other. Both of them have shortcomings.

The two-layer sharp-interface theory consists of two coupled fourth-order partial differential equations that describe the evolution in time of the thickness profiles of the two layers of liquid. The liquids are assumed to be entirely immiscible. This is a good approximation for strong segregation between the two liquids. The model is not able to describe the formation of depletion/enrichment layers near the liquid-solid and liquid-gas interfaces. However, one may easily incorporate the various short- and long-range wetting interactions between the four involved media and study dewetting and coarsening pathways (Pototsky et al., 2005, 2006). See also the alternative model by Bandyopadhyay et al. (2005) extended to incorporate the influence of an electrical field by Bandyopadhyay et al. (2009). Note that the system of evolution equations represents the general form of coupled evolution equations for two conserved fields in gradient dynamics based on a underlying energy functional. With other words, Eqs. (9) form the natural extension of Eq. (3). This makes it straightforward to incorporate additional interactions solely through amendments of the energy functional.

Whereas for the two-layer sharp-interface theory the restriction to the evolution of a layered film is part of the model, this is not the case for the diffuse-interface short-wave model discussed in Section 4. There the restriction results from our chosen method of analysis, i.e., a combination of studies of steady layered states and of their linear stability with respect to lateral perturbations. In principle, one can study the full time evolution of homogeneous films numerically and check for which parameter values our assumption holds. This has not yet been done. One may also use the static limit of the model to analyse the final states, i.e., films with modulated surface and concentration profiles (see Frastia et al. (2011b) and Bribesh et al. (2011)). A model that is simpler in terms of boundary conditions is studied by Frolovskaya et al. (2008). A drawback of the usage of model-H with boundary conditions at the substrate and free film surface is that it does up to now not account for the wettability of the mixture itself. Wettability effects are, however, fully incorporated in the two-layer sharp interface approach (within the restrictions posed by the long-wave approximation). The drawback can be seen by considering a stable mixture without surface biases. The model does not contain any information on the wettability of the mixture on the substrate like, e.g., a disjoining pressure and could not model the formation of a drop of mixture on a 'dry' substrate in the case of partial wetting. It does, however, contain a partial information on the 'relative wettability' of the two components with respect to the gas above the film and the substrate. This aspect enters the model via the energetic biases at the free surface and the substrate, respectively. The bias results in enrichment/depletion layers that may for thin enough layers interact with the diffuse liquid-liquid interface. This is discussed by Pismen and Pomeau (2000) for a diffuse interface description of a film of simple liquid on a solid substrate. They show that a long-wave approximation of the diffuse interface description coupled to hydrodynamics (model-H) results in a thin film equation as our Eq. (1). However the disjoining pressure it contains results from the interaction of a depletion layer at the solid substrate (resulting from partial wetting) and the diffuse liquid-gas interface. For more details see Pismen and Pomeau (2000).

An alternative approach to the problem of decomposing and dewetting mixture is proposed by Clarke (2004, 2005). There they propose and analyse a simple model based on phenomenological non-equilibrium thermodynamics in a gradient dynamics setting. Assuming that concentration depends only on the lateral coordinate the author states coupled evolution equations for the concentration and height fields (Clarke, 2005). We think, however, that their system of equations needs further discussion in the context of other long-wave models for thin films of solutions and suspensions (Warner et al., 2003; Cook et al., 2008; Thiele et al., 2009b; Frastia et al., 2011a). It is not obvious to us how the limit of passive solute (i.e., no decomposition) of Clarke (2005) has to be taken to obtain the evolution equation of, e.g., Warner et al. (2003) (in their case without surfactant). Recently, another long-wave model has been derived *from model-H* under the similar assumption that concentration depends only on the lateral coordinate (Náraigh and Thiffeault, 2010). For a passive solute it correctly reduces to the corresponding cases of Warner et al. (2003); Cook et al. (2008); Frastia et al. (2011a). It is shown in Thiele (2011b) that the model derived in Náraigh and Thiffeault (2010) can be written as a gradient dynamics for two coupled conserved fields in the same form as Eqs. (9)[1].The resulting evolution equations lend themselves nicely to extensions as, e.g., the incorporation of a concentration-dependent wettability. Replacing $g(\phi)$ by a purely entropic contribution and dropping the gradient term in the concentration one recovers the usual thin film model for solutions and suspensions (Warner et al., 2003; Frastia et al., 2011a).

With the present contribution we have aimed at providing a description of a number of continuum approaches to thin films of liquid mixtures as an example of a thin film of complex liquids. Other important and often more complex settings are increasingly investigated experimentally and theoretically. Examples include films of (evaporating) solutions of polymers or nanoparticles and their mixtures, films covered by soluble or insoluble, low- or high-density surfactants, and films of nematic liquid crystals or active liquids. The involved processes and effects include convective and diffusive motion, phase separation and other phase transitions, evaporation/condensation of solute and complex rheologies, capillarity and wettability. For many of the intricate phenomena only partial theoretical descriptions exist. As starting points for further studies we refer the reader to reviews by Matar and Craster (2009) and Bonn et al. (2009), and recent particular studies like, e.g., Cummings (2004); Sankararaman and Ramaswamy (2009); Náraigh and Thiffeault (2010); Köpf et al. (2010); Frastia et al. (2011a); Thiele (2011b) and references therein.

[1]Replace h_1 and h_2 in Eqs. (9) by film thickness $h(x,t)$ and the local amount of solute (effective local solute layer thickness) $\psi(x,t) = h(x,t)\,\phi(x,t)$, respectively, where ϕ is a (non-conserved) vertically averaged concentration. Therefore, h and ψ are both conserved fields. The corresponding mobilities are (here non-dimensional)

$$\mathbf{Q} = \begin{pmatrix} Q_{hh} & Q_{h\psi} \\ Q_{\psi h} & Q_{\psi\psi} \end{pmatrix} = \frac{1}{3}\begin{pmatrix} h^3 & h^2\psi \\ h^2\psi & h\psi^2 + 3\widehat{D}\psi \end{pmatrix} \tag{22}$$

where $\widehat{D}$ is a diffusion number and the energy functional is $F[h,\psi] = \int \left[\frac{1}{2}(\nabla h)^2 + \hat{\sigma}(\nabla(\psi/h))^2 + f(h) + h\,g\left(\frac{\psi}{h}\right)\right] dA$ with $g(\phi) \sim (\phi^2-1)^2$. For details see Thiele (2011b).

Acknowledgments

The author acknowledges all collegues who have contributed to the works that form the basis of the present chapter. In particular, some of the original figures were produced by A. Pototsky and S. Madruga. Recent discussions with A. J. Archer, F. Bribesh, L. Frastia, M. Galvagno, L. M. Pismen, M. Plapp, and D. Todorova have influenced the writing. The work was in part supported by the European Union via the FP7 Marie Curie scheme [Grant PITN-GA-2008-214919 (MULTIFLOW)], and the Ecole Polytechnique (Paris) where part of the text was written. Finally, I thank R. Mauri for the organisation of the Summer School in Udine where the material was originally taught. It was further developed for Summer Schools of the ITN MULTIFLOW in Paris (2009) and El Escorial (2010).

Bibliography

V. S. Ajaev. Spreading of thin volatile liquid droplets on uniformly heated surfaces. J. Fluid Mech., 528:279–296, 2005.

D. M. Anderson, G. B. McFadden, and A. A. Wheeler. Diffuse-interface methods in fluid mechanics. Ann. Rev. Fluid Mech., 30:139–165, 1998. doi: 10.1146/annurev.fluid.30.1.139.

A. J. Archer, M. J. Robbins, and U. Thiele. Dynamical density functional theory for the dewetting of evaporating thin films of nanoparticle suspensions exhibiting pattern formation. Phys. Rev. E, 81(2):021602, 2010. doi: 10.1103/PhysRevE.81.021602.

D. Bandyopadhyay, R. Gulabani, and A. Sharma. Stability and dynamics of bilayers. Ind. Eng. Chem. Res., 44:1259–1272, 2005.

D. Bandyopadhyay, A. Sharma, U. Thiele, and P. D. S. Reddy. Electric field induced interfacial instabilities and morphologies of thin viscous and elastic bilayers. Langmuir, 25:9108–9118, 2009. doi: 10.1021/la900635f.

F. R. S. Batchelor. An Introduction to Fluid Dynamics. University Press, Cambridge, 2000.

J. Becker, G. Grün, R. Seemann, H. Mantz, K. Jacobs, K. R. Mecke, and R. Blossey. Complex dewetting scenarios captured by thin-film models. Nat. Mater., 2:59–63, 2003.

P. Beltrame and U. Thiele. Time integration and steady-state continuation method for lubrication equations. SIAM J. Appl. Dyn. Syst., 9:484–518, 2010. doi: 10.1137/080718619.

P. Beltrame, P. Hänggi, and U. Thiele. Depinning of three-dimensional drops from wettability defects. Europhys. Lett., 86:24006, 2009. doi: 10.1209/0295-5075/86/24006.

P. Beltrame, E. Knobloch, P. Hänggi, and U. Thiele. Rayleigh and depinning instabilities of forced liquid ridges on heterogeneous substrates. Phys. Rev. E, 83:016305, 2011. doi: 10.1103/PhysRevE.83.016305.

D. J. Benney. Long waves on liquid films. J. Math. & Phys., 45:150–155, 1966.

M. Bestehorn and K. Neuffer. Surface patterns of laterally extended thin liquid films in three dimensions. Phys. Rev. Lett., 87:046101, 2001. doi: 10.1103/PhysRevLett.87.046101.

M. Böltau, S. Walheim, J. Mlynek, G. Krausch, and U. Steiner. Surface-induced structure formation of polymer blends on patterned substrates. Nature, 391:877–879, 1998.

D. Bonn, J. Eggers, J. Indekeu, J. Meunier, and E. Rolley. Wetting and spreading. Rev. Mod. Phys., 81:739–805, 2009. doi: 10.1103/RevModPhys.81.739.

W. Boos and A. Thess. Cascade of structures in long-wavelength Marangoni instability. Phys. Fluids, 11:1484–1494, 1999.

M. S. Borgas and J. B. Grotberg. Monolayer flow on a thin film (lung application). J. Fluid Mech., 193:151–170, 1988.

F. Bribesh, L. Frastia, and U. Thiele. 2011. (in preparation).

M. Brinkmann and R. Lipowsky. Wetting morphologies on substrates with striped surface domains. J. Appl. Phys., 92:4296–4306, 2002.

J. M. Burgess, A. Juel, W. D. McCormick, J. B. Swift, and H. L. Swinney. Suppression of dripping from a ceiling. Phys. Rev. Lett., 86:1203–1206, 2001.

J. W. Cahn. Phase separation by spinodal decomposition in isotropic systems. J. Chem. Phys., 42:93–99, 1965.

N. Clarke. Instabilities in thin-film binary mixtures. Eur. Phys. J. E, 14: 207–210, 2004.

N. Clarke. Toward a model for pattern formation in ultrathin-film binary mixtures. Macromolecules, 38:6775–6778, 2005.

B. P. Cook, A. L. Bertozzi, and A. E. Hosoi. Shock solutions for particle-laden thin films. SIAM J. Appl. Math., 68:760–783, 2008. doi: 10.1137/060677811.

R. V. Craster and O. K. Matar. Dynamics and stability of thin liquid films. Rev. Mod. Phys., 81:1131–1198, 2009. doi: 10.1103/RevModPhys.81.1131.

L. J. Cummings. Evolution of a thin film of nematic liquid crystal with anisotropic surface energy. Eur. J. Appl. Math., 15:651–677, 2004.

M. O. David, G. Reiter, T. Sitthai, and J. Schultz. Deformation of a glassy polymer film by long-range intermolecular forces. Langmuir, 14:5667–5672, 1998.

J. De Coninck and T. D. Blake. Wetting and molecular dynamics simulations of simple liquids. Ann. Rev. Mater. Res., 38:1–22, 2008. doi: 10.1146/annurev.matsci.38.060407.130339.

P.-G. de Gennes. Wetting: Statics and dynamics. Rev. Mod. Phys., 57: 827–863, 1985. doi: 10.1103/RevModPhys.57.827.

P.-G. de Gennes. The dynamics of reactive wetting on solid surfaces. Physica A, 249:196–205, 1998.

G. J. Dunn, S. K. Wilson, B. R. Duffy, S. David, and K. Sefiane. The strong influence of substrate conductivity on droplet evaporation. J. Fluid Mech., 623:329–351, 2009. doi: 10.1017/S0022112008005004.

I. E. Dzyaloshinskii, E. M. Lifshitz, and L. P. Pitaevskii. Van der Waals forces in liquid films. Sov. Phys. JETP, 37:161, 1960.

H. P. Fischer, P. Maass, and W. Dieterich. Diverging time and length scales of spinodal decomposition modes in thin films. Europhys. Lett., 42:49–54, 1998.

L. Frastia, A. J. Archer, and U. Thiele. Dynamical model for the formation of patterned deposits at receding contact lines. Phys. Rev. Lett., 2011a. at press, (preprint at http://arxiv.org/abs/1008.4334).

L. Frastia, U. Thiele, and L. M. Pismen. Determination of the thickness and composition profiles for a film of binary mixture on a solid substrate. Math. Model. Nat. Phenom., 6:62–86, 2011b. doi: 10.1051/mmnp/20116104.

O. A. Frolovskaya, A. A. Nepomnyashchy, A. Oron, and A. A. Golovin. Stability of a two-layer binary-fluid system with a diffuse interface. Phys. Fluids, 20:112105, 2008. doi: 10.1063/1.3021479.

D. Gallez and W. T. Coakley. Far-from-equilibrium phenomena in bioadhesion processes. Heterogeneous Chem. Rev., 3:443–475, 1996.

M. Geoghegan and G. Krausch. Wetting at polymer surfaces and interfaces. Prog. Polym. Sci., 28:261–302, 2003. doi: 10.1016/S0079-6700(02)00080-1.

A. A. Golovin, A. A. Nepomnyashchy, S. H. Davis, and M. A. Zaks. Convective Cahn-Hilliard models: From coarsening to roughening. Phys. Rev. Lett., 86:1550–1553, 2001. doi: 10.1103/PhysRevLett.86.1550.

P. C. Hohenberg and B. I. Halperin. Theory of dynamic critical phenomena. Rev. Mod. Phys., 49:435–479, 1977.

J. Israelachvili. Intermolecular and Surface Forces. Academic Press: London, 1992.

O. E. Jensen and J. B. Grotberg. Insoluble surfactant spreading on a thin viscous film: Shock evolution and film rupture. J. Fluid Mech., 240: 259–288, 1992.

K. John and U. Thiele. Self-ratcheting stokes drops driven by oblique vibrations. Phys. Rev. Lett., 104:107801, 2010. doi: 10.1103/PhysRevLett.104.107801.

K. John, M. Bär, and U. Thiele. Self-propelled running droplets on solid substrates driven by chemical reactions. Eur. Phys. J. E, 18:183–199, 2005. doi: 10.1140/epje/i2005-10039-1.

K. John, P. Hänggi, and U. Thiele. Ratchet-driven fluid transport in bounded two-layer films of immiscible liquids. Soft Matter, 4:1183–1195, 2008. doi: 10.1039/b718850a.

D. D. Joseph. Fluid-dynamics of 2 miscible liquids with diffusion and gradient stresses. Eur. J. Mech. B-Fluids, 9:565–596, 1990.

S. Kalliadasis. Falling films under complicated conditions. In S. Kalliadasis and U. Thiele, editors, Thin films of Soft Matter, pages 137–190, Wien, 2007. Springer.

R. Konnur, K. Kargupta, and A. Sharma. Instability and morphology of thin liquid films on chemically heterogeneous substrates. Phys. Rev. Lett., 84:931–934, 2000. doi: 10.1103/PhysRevLett.84.931.

M. H. Köpf, S. V. Gurevich, R. Friedrich, and L. F. Chi. Pattern formation in monolayer transfer systems with substrate-mediated condensation. Langmuir, 26:10444–10447, 2010. doi: 10.1021/la101900z.

S. Krishnamoorthy, B. Ramaswamy, and S. W. Joo. Spontaneous rupture of thin liquid films due to thermocapillarity: A full-scale direct numerical simulation. Phys. Fluids, 7:2291–2293, 1995.

J. S. Langer. An introduction to the kinetics of first-order phase transitions. In C. Godreche, editor, Solids far from Equilibrium, pages 297–363. Cambridge University Press, 1992.

F. Léonforte, J. Servantie, C. Pastorino, and M. Müller. Molecular transport and flow past hard and soft surfaces: computer simulation of model systems. J. Phys.: Cond. Mat., 2011. (at press).

Z. Lin, T. Kerle, T. P. Russell, E. Schäffer, and U. Steiner. Structure formation at the interface of liquid liquid bilayer in electric field. Macromolecules, 35:3971–3976, 2002.

S. Lindström and H. Andersson-Svahn. Miniaturization of biological assays – Overview on microwell devices for single-cell analyses. Biochim. Biophys. Acta, 2010. doi: 10.1016/j.bbagen.2010.04.009. published online.

A. V. Lyushnin, A. A. Golovin, and L. M. Pismen. Fingering instability of thin evaporating liquid films. Phys. Rev. E, 65:021602, 2002. doi: 10.1103/PhysRevE.65.021602.

S. Madruga and U. Thiele. Decomposition driven interface evolution for layers of binary mixtures: II. Influence of convective transport on linear stability. Phys. Fluids, 21:062104, 2009. doi: 10.1063/1.3132789.

O. K. Matar and R. V. Craster. Dynamics of surfactant-assisted spreading. Soft Matter, 5:3801–3809, 2009. doi: 10.1039/b908719m.

D. Merkt, A. Pototsky, M. Bestehorn, and U. Thiele. Long-wave theory of bounded two-layer films with a free liquid-liquid interface: Short- and long-time evolution. Phys. Fluids, 17:064104, 2005. doi: 10.1063/1.1935487.

M. Mertig, U. Thiele, J. Bradt, G. Leibiger, W. Pompe, and H. Wendrock. Scanning force microscopy and geometrical analysis of two-dimensional collagen network formation. Surface and Interface Analysis, 25:514–521, 1997.

D. Mijatovic, J. C. T. Eijkel, and A. van den Berg. Technologies for nanofluidic systems: Top-down vs. bottom-up - a review. Lab Chip, 5:492–500, 2005.

V. S. Mitlin. Dewetting of solid surface: Analogy with spinodal decomposition. J. Colloid Interface Sci., 156:491–497, 1993. doi: 10.1006/jcis.1993.1142.

M. D. Morariu, N. E. Voicu, E. Schäffer, Z. Lin, T. P. Russell, and U. Steiner. Hierarchical structure formation and pattern replication induced by an electric field. Nat. Mater., 2:48–52, 2003. doi: 10.1038/nmat789.

A. Münch, B. Wagner, and T. P. Witelski. Lubrication models with small to large slip lengths. J. Eng. Math., 53:359–383, 2005. doi: 10.1007/s10665-005-9020-3.

L. Ó. Náraigh and J. L. Thiffeault. Nonlinear dynamics of phase separation in thin films. Nonlinearity, 23:1559–1583, 2010. doi: 10.1088/0951-7715/23/7/003.

A. Oron and P. Rosenau. Formation of patterns induced by thermocapillarity and gravity. J. Physique II France, 2:131–146, 1992.

A. Oron, S. H. Davis, and S. G. Bankoff. Long-scale evolution of thin liquid films. Rev. Mod. Phys., 69:931–980, 1997. doi: 10.1103/RevModPhys.69.931.

M. Oron, T. Kerle, R. Yerushalmi-Rozen, and J. Klein. Persistent droplet motion in liquid-liquid dewetting. Phys. Rev. Lett., 92:236104, 2004. doi: 10.1103/PhysRevLett.92.236104.

Q. Pan, K. I. Winey, H. H. Hu, and R. J. Composto. Unstable polymer bilayers. 2. The effect of film thickness. Langmuir, 13:1758–1766, 1997.

A. Z. Panagiotopoulos. Monte Carlo methods for phase equilibria of fluids. J. Phys.: Condens. Matter, 12:R25–R52, 2000.

E. Pauliac-Vaujour, A. Stannard, C. P. Martin, M. O. Blunt, I. Notingher, P. J. Moriarty, I. Vancea, and U. Thiele. Fingering instabilities in dewetting nanofluids. Phys. Rev. Lett., 100:176102, 2008. doi: 10.1103/PhysRevLett.100.176102.

A. Pereira, P. M. J. Trevelyan, U. Thiele, and S. Kalliadasis. Interfacial hydrodynamic waves driven by chemical reactions. J. Engg. Math., 59: 207–220, 2007. doi: 10.1007/s10665-007-9143-9.

T. Pfohl, F. Mugele, R. Seemann, and S. Herminghaus. Trends in microfluidics with complex fluids. ChemPhysChem, 4:1291–1298, 2003.

L. M. Pismen. Nonlocal diffuse interface theory of thin films and the moving contact line. Phys. Rev. E, 64:021603, 2001. doi: 10.1103/PhysRevE.64.021603.

L. M. Pismen. Mesoscopic hydrodynamics of contact line motion. Colloid Surf. A-Physicochem. Eng. Asp., 206:11–30, 2002.

L. M. Pismen and Y. Pomeau. Disjoining potential and spreading of thin liquid layers in the diffuse interface model coupled to hydrodynamics. Phys. Rev. E, 62:2480–2492, 2000. doi: 10.1103/PhysRevE.62.2480.

M. Plapp and J. F. Gouyet. Surface modes and ordered patterns during spinodal decomposition of an abv model alloy. Phys. Rev. Lett., 78: 4970–4973, 1997.

A. Pototsky, M. Bestehorn, D. Merkt, and U. Thiele. Morphology changes in the evolution of liquid two-layer films. J. Chem. Phys., 122:224711, 2005. doi: 10.1063/1.1927512.

A. Pototsky, M. Bestehorn, D. Merkt, and U. Thiele. 3d surface patterns in liquid two-layer films. Europhys. Lett., 74:665–671, 2006. doi: 10.1209/epl/i2006-10026-8.

E. Rabani, D. R. Reichman, P. L. Geissler, and L. E. Brus. Drying-mediated self-assembly of nanoparticles. Nature, 426:271–274, 2003.

A. Y. Rednikov and P. Colinet. Vapor-liquid steady meniscus at a superheated wall: Asymptotics in an intermediate zone near the contact line. Microgravity Sci. Technol., 22:249–255, 2010. doi: 10.1007/s12217-010-9177-x.

G. Reiter. Dewetting of thin polymer films. Phys. Rev. Lett., 68:75–78, 1992. doi: 10.1103/PhysRevLett.68.75.

G. Reiter and A. Sharma. Auto-optimization of dewetting rates by rim instabilities in slipping polymer films. Phys. Rev. Lett., 87:166103, 2001. doi: 10.1103/PhysRevLett.87.166103.

D. H. Rothman and S. Zaleski. Lattice-gas models of phase-separation - interfaces, phase-transitions, and multiphase flow. Rev. Mod. Phys., 66: 1417–1479, 1994.

E. Ruckenstein and R. K. Jain. Spontaneous rupture of thin liquid films. J. Chem. Soc. Faraday Trans. II, 70:132–147, 1974.

C. Ruyer-Quil and P. Manneville. Modeling film flows down inclined planes. Eur. Phys. J. B, 6:277–292, 1998.

A. J. Ryan, C. J. Crook, J. R. Howse, P. Topham, R. A. L. Jones, M. Geoghegan, A. J. Parnell, L. Ruiz-Perez, S. J. Martin, A. Cadby, A. Menelle, J. R. P. Webster, A. J. Gleeson, and W. Bras. Responsive brushes and gels as components of soft nanotechnology. Faraday Discuss., 128:55–74, 2005.

I. M. R. Sadiq and R. Usha. Thin Newtonian film flow down a porous inclined plane: Stability analysis. Phys. Fluids, 20:022105, 2008.

S. Sankararaman and S. Ramaswamy. Instabilities and waves in thin films of living fluids. Phys. Rev. Lett., 102:118107, 2009. doi: 10.1103/PhysRevLett.102.118107.

B. Scheid, C. Ruyer-Quil, U. Thiele, O. A. Kabov, J. C. Legros, and P. Colinet. Validity domain of the Benney equation including Marangoni effect for closed and open flows. J. Fluid Mech., 527:303–335, 2005. doi: 10.1017/S0022112004003179.

R. Seemann, S. Herminghaus, C. Neto, S. Schlagowski, D. Podzimek, R. Konrad, H. Mantz, and K. Jacobs. Dynamics and structure formation in thin polymer melt films. J. Phys.: Condens. Matter, 17:S267–S290, 2005.

M. Sferrazza, M. Heppenstall-Butler, R. Cubitt, D. Bucknall, J. Webster, and R. A. L. Jones. Interfacial instability driven by dispersive forces: The early stages of spinodal dewetting of a thin polymer film on a polymer substrate. Phys. Rev. Lett., 81:5173–5176, 1998.

A. Sharma. Relationship of thin film stability and morphology to macroscopic parameters of wetting in the apolar and polar systems. Langmuir, 9:861–869, 1993a. doi: 10.1021/la00027a042.

A. Sharma. Equilibrium contact angles and film thicknesses in the apolar and polar systems: Role of intermolecular interactions in coexistence of drops with thin films. Langmuir, 9:3580, 1993b.

A. Sharma and R. Khanna. Pattern formation in unstable thin liquid films. Phys. Rev. Lett., 81:3463–3466, 1998. doi: 10.1103/PhysRevLett.81.3463.

A. Sharma and G. Reiter. Instability of thin polymer films on coated substrates: Rupture, dewetting and drop formation. J. Colloid Interface Sci., 178:383–399, 1996. doi: 10.1006/jcis.1996.0133.

A. Sharma and E. Ruckenstein. Mechanism of tear film rupture and its implications for contact-lens tolerance. Amer. J. Optom. Physiol. Opt., 62:246–253, 1985.

T. M. Squires and S. R. Quake. Microfluidics: Fluid physics at the nanoliter scale. Rev. Mod. Phys., 77:977–1026, 2005.

V. M. Starov and M. G. Velarde. Surface forces and wetting phenomena. J. Phys.-Condes. Matter, 21:464121, 2009. doi: 10.1088/0953-8984/21/46/464121.

U. Thiele. Open questions and promising new fields in dewetting. Eur. Phys. J. E, 12:409–416, 2003. doi: 10.1140/epje/e2004-00009-4.

U. Thiele. Structure formation in thin liquid films. In S. Kalliadasis and U. Thiele, editors, Thin films of Soft Matter, pages 25–93, Wien, 2007. Springer.

U. Thiele. Thin film evolution equations from (evaporating) dewetting liquid layers to epitaxial growth. J. Phys.: Condens. Matter, 22:084019, 2010. doi: 10.1088/0953-8984/22/8/084019.

U. Thiele. On the depinning of a drop of partially wetting liquid on a rotating cylinder. J. Fluid Mech., 671, 121–136, 2011a. doi: 10.1017/S0022112010005483.

U. Thiele. Note on thin film equations for solutions and suspensions. Eur. Phys. J. Special Topics, 197:213–220, 2011b. doi: 10.1140/epjst/e2011-01462-7.

U. Thiele. Thoughts on mesoscopic continuum models. Eur. Phys. J. Special Topics, 197:67-71, 2011c. doi: 10.1140/epjst/e2011-01438-7.

U. Thiele and K. John. Transport of free surface liquid films and drops by external ratchets and self-ratcheting mechanisms. Chem. Phys., 375: 578–586, 2010. doi: 10.1016/j.chemphys.2010.07.011.

U. Thiele and E. Knobloch. Thin liquid films on a slightly inclined heated plate. Physica D, 190:213–248, 2004.

U. Thiele, M. G. Velarde, and K. Neuffer. Dewetting: Film rupture by nucleation in the spinodal regime. Phys. Rev. Lett., 87:016104, 2001a. doi: 10.1103/PhysRevLett.87.016104.

U. Thiele, M. G. Velarde, K. Neuffer, M. Bestehorn, and Y. Pomeau. Sliding drops in the diffuse interface model coupled to hydrodynamics. Phys. Rev. E, 64:061601, 2001b. doi: 10.1103/PhysRevE.64.061601.

U. Thiele, K. Neuffer, Y. Pomeau, and M. G. Velarde. On the importance of nucleation solutions for the rupture of thin liquid films. Colloid Surf. A, 206:135–155, 2002.

U. Thiele, L. Brusch, M. Bestehorn, and M. Bär. Modelling thin-film dewetting on structured substrates and templates: Bifurcation analysis and numerical simulations. Eur. Phys. J. E, 11:255–271, 2003. doi: 10.1140/epje/i2003-10019-5.

U. Thiele, S. Madruga, and L. Frastia. Decomposition driven interface evolution for layers of binary mixtures: I. Model derivation and stratified base states. Phys. Fluids, 19:122106, 2007. doi: 10.1063/1.2824404.

U. Thiele, B. Goyeau, and M. G. Velarde. Film flow on a porous substrate. Phys. Fluids, 21:014103, 2009a. doi: 10.1063/1.3054157.

U. Thiele, I. Vancea, A. J. Archer, M. J. Robbins, L. Frastia, A. Stannard, E. Pauliac-Vaujour, C. P. Martin, M. O. Blunt, and P. J. Moriarty. Modelling approaches to the dewetting of evaporating thin films of nanoparticle suspensions. J. Phys.-Cond. Mat., 21:264016, 2009b. doi: 10.1088/0953-8984/21/26/264016.

D. Todorova, U. Thiele, and L. M. Pismen. The relation of steady evaporating drops fed by an influx and freely evaporating drops. J. Engg. Math., 2011. doi: 10.1007/s10665-011-9485-1 (online).

I. Vancea, U. Thiele, E. Pauliac-Vaujour, A. Stannard, C. P. Martin, M. O. Blunt, and P. J. Moriarty. Front instabilities in evaporatively dewetting nanofluids. Phys. Rev. E, 78:041601, 2008. doi: 10.1103/PhysRevE.78.041601.

R. Verma and A. Sharma. Defect sensitivity in instability and dewetting of thin liquid films: Two regimes of spinodal dewetting. Ind. Eng. Chem. Res., 46:3108–3118, 2007. doi: 10.1021/ie060615q.

R. Verma, A. Sharma, K. Kargupta, and J. Bhaumik. Electric field induced instability and pattern formation in thin liquid films. Langmuir, 21: 3710–3721, 2005. doi: 10.1021/la0472100.

N. Vladimirova, A. Malagoli, and R. Mauri. Two-dimensional model of phase segregation in liquid binary mixtures. Phys. Rev. E, 60:6968–6977, 1999.

M. R. E. Warner, R. V. Craster, and O. K. Matar. Surface patterning via evaporation of ultrathin films containing nanoparticles. J. Colloid Interface Sci., 267:92–110, 2003.

J. Xu, J. F. Xia, and Z. Q. Lin. Evaporation-induced self-assembly of nanoparticles from a sphere-on-flat geometry. Angew. Chem.-Int. Edit., 46:1860–1863, 2007. doi: 10.1002/anie.200604540.

Phase-Field Models

Mathis Plapp
Physique de la Matière Condensée, École Polytechnique, CNRS,
91128 Palaiseau, France

Abstract. Phase-field models have become popular in recent years to describe a host of free-boundary problems in various areas of research. The key point of the phase-field approach is that surfaces and interfaces are implicitly described by continuous scalar fields that take constant values in the bulk phases and vary continuously but steeply across a diffuse front. In the present contribution, a distinction is made between models in which the phase field can be identified with a physical quantity (coarse-grained on a mesoscopic scale), and models in which the phase field can only be interpreted as a smoothed indicator function. Simple diffuse-interface models for the motion of magnetic domain walls, the growth of precipitates in binary alloys, and for solidification are reviewed, and it is pointed out that is such models the free energy function determines both the bulk behavior of the dynamic variable and the properties of the interface. Next, a phenomenological phase-field model for solidification is introduced, and it is shown that with a proper choice of some interpolation functions, surface and bulk properties can be adjusted independently in this model. The link between this phase-field model and the classic free-boundary formulation of solidification is established by the use of matched asymptotic analysis. The results of this analysis can then be exploited to design new phase-field models that cannot be derived by the standard variational procedure from simple free energy functionals within the thermodynamic framework. As examples for applications of this approach, the solidification of alloys and the advected field model for two-phase flow are briefly discussed.

1 Introduction

Phase-field models have rapidly gained popularity over the last two decades in various fields. Some examples for their applications are discussed in recent reviews on the formation of microstructures during solidification by Boettinger et al. (2002) and Plapp (2007), on solid-state transformations by Chen (2002),Steinbach (2009), and Wang and Li (2010), and on multiphase flows by Anderson et al. (1998), but it is almost impossible to give an exhaustive list of the topics treated with the help of phase-field methods since the development of phase-field models for ever new applications is a rapidly advancing field. All the topics mentioned above have in common that they involve the motion of interfaces or boundaries in response to a coupling of the boundary with one or several transport fields (such as diffusion, flow, stress or temperature fields). This interaction generates morphological instabilities and leads to the spontaneous emergence of complex structures.

All these different models have in common that they describe the geometry of the boundaries through one or several *phase fields*. This name was originally coined in solidification, where the phase field indicates in which thermodynamic phase (solid or liquid) each point of space is located. These fields have fixed pretedermined values in each of the domains occupied by a bulk phase, and vary continuously from one bulk value to the other through an interface that has a well-defined width. In other words, in all phase-field models the interfaces are *diffuse*.

In fact, in the literature, the terms "diffuse-interface model" and "phase-field model" are often used as synonyms. In contrast, in the present contribution I will make a distinction between these two classes of models that is based on the two different and complementary viewpoints that can be taken to derive them. In the first perspective, which could be called "bottom-up", one starts from a microscopic picture of a physical system and performs a *coarse-graining*. While this operation can rarely be carried out explicitly, conceptually it is well defined. As an example, consider a liquid-vapor interface. The relevant quantity that characterizes the difference between the two phases is the number density of molecules, which is high in the liquid but low in the vapor. However, to define a smooth density, the local number of atoms has to be averaged over a volume that is large enough to contain a significant number of molecules, but small enough to remain "local", that is, smaller than any geometric scale of the two-phase pattern to be described. Then, the total free energy of the system may be expressed as a functional that is obtained by averaging all quantities on the coarse-graining scale, and all the coefficients of the functional can in principle be calculated from the elementary intermolecular interactions. The first example for such a model was the van der Waals theory of the liquid-vapor interface, see Rowlinson (1979), but since then many other similar models have been developed, for instance for magnetic domain walls (where the local variable is the magnetization) or interfaces between two phases in a binary mixture (where the relevant variable is the composition). All of these models exhibit diffuse interfaces, whose characteristic thickness agrees with the actual intrinsic thickness of the physical interfaces, which are indeed microscopically rough and therefore diffuse.

The opposite point of view could be called "top-down". It exploits the scale separation that is inherent to most of the systems mentioned above. Indeed, consider a mixture of two immiscible fluids. The domains occupied by the two fluids can be of any size, but in flows under ordinary conditions droplets are typically of millimetric size. In contrast, the width of the diffuse interfaces between the two fluids is of the order of the molecular size, that is Angstroms for simple fluids and nanometers for polymer mixtures. Therefore, the interface thickness is several orders of magnitude smaller than any geometric length scale of the flow pattern. This is the reason why such problems have been very successfully described as *free boundary problems*: the interfaces are assumed to be mathematically sharp, all quantities that differ between the two phases in contact exhibit jumps at the interfaces, and the transport equations in the bulk phases are solved with explicit boundary conditions imposed at the surface. Capillary forces created by the surface tension are localized at the sharp interfaces and must thus be described by Dirac distributions. This sharp-interface formulation can be perfectly used (and has been used) to perform numerical simulations. However, the handling of the interfaces (which must be discretized in some way) is cumbersome. An obvious idea to make this formulation more amenable to numerical treatment is to smooth out the singularities, that is, to replace step functions and surface delta functions by continuous profiles that have the shape of a smooth kink and a smooth peak, respectively. In this picture, the phase field is just the regularized step (or indicator) function. This regularization of singularities is a mathematically well-defined procedure, which introduces a new length scale into the problem, namely, the typical thickness W of the kink solution, which is *a priori* a free parameter. Quite naturally, the resulting regularized problem will behave differently from the original singular problem. However, the differences disappear in the so-called sharp-interface limit, in which the thickness of the interface tends to zero while all the macroscopic scales remain fixed. A more sophisticated use of this procedure is the so-called thin-interface limit, in which the first-order corrections to the macroscopic problem are calculated and taken into account when choosing the parameters of the phase-field model. In that case, the first corrections to the sharp-interface problem scale as W^2, such that precise simulations can be achieved with much larger values of W. Such models have been termed "quantitative" in the literature; see Karma and Rappel (1996, 1998); Echebarria et al. (2004). Obviously, an upper limit for W is set by the other macroscopic scales present in the problem. Indeed, W must remain much smaller (in practice, about one order of magnitude) than any geometric length of the two-phase pattern in order that the smoothened version of the problem remains a correct description of the macroscopic geometry.

In the following, I will refer to the models in which the interface is decribed by the profile of a physically accessible quantity as "diffuse-interface models", whereas I will designate by the term "phase-field models" the models in which the interface is described by a smoothed indicator function. While this is not a terminology that is standard in the literature, I believe that it is highly useful to distinghish the two philosophies in order to gain a thorough understanding of the foundations of such models.

For the development of a successful model, it is often advantageous to use both points of view. Indeed, if a new physical situation is considered, it is often quite easy to identify the suitable order parameters together with their symmetries, which gives precious hints for writing down a model that is physically correct. Furthermore, standard out-of-equilibrium thermodynamics then directly gives the equations of motion through a variational procedure. However, models obtained in this way are rarely useful for practical applications, because of the separation of length scales already discussed above. If the diffuse interfaces have the same thickness as the physical ones, the equations need to be discretized on this microscopic scale to correctly resolve the interface profiles. This implies that the length and time scales that can be attained in numerical simulations are severely limited. The only way to access larger system sizes and longer time scales is to use interfaces that are much thicker than the physical ones. However, if the interface thickness is upscaled in a diffuse-interface model, physical effects explicitly linked to the finite interface thickness will generally be amplified, which makes simulation results unreliable. In phase-field models, the separation between the "interface marker" (the phase field) and the directly accessible physical quantities makes it easier to modify the equations in such a way that this amplification of interface effects is avoided.

In the remainder of this article, I will develop several examples which illustrate the relations and distinctions between diffuse-interface and phase-field models, and which demonstrate that phase-field models can be used for obtaining extremely precise results with robust and simple numerical algorithms. In the next section, I will introduce several diffuse-interface models to illustrate some of their properties. In section 3, I will introduce the phase-field model for the solificiation of a pure substance, discuss the difference with diffuse-interface models, and then discuss a method to relate the model parameters to physical quantities for arbitrary interface thickness. Due to space limitations, I will not explicitly display examples the demonstrate the performance of this model, but I will give detailed references to relevant literature where more details can be found. In section 4, two examples will be given that show how the results of the asymptotic analysis can be used to remove thin-interface effects from phase-field models by adding new terms to the equations that sometimes cannot be derived from free energy functionals within a variational framework. Brief conclusions are given in section 5.

2 Simple diffuse-interface models

In this section, I will introduce some of the simplest diffuse-interface models for subsequent comparison with phase-field models. The presentation will be short and limited to the material that is relevant for the present discussion; more details can be found in the review articles by Hohenberg and Halperin (1977) and Bray (1994), and a particularly pedagogic introduction to some of these models is given by Langer (1991).

2.1 Ising ferromagnet

The Ising model is one of the simplest statistical mechanics models: on a regular lattice, each site is occupied by a microscopic spin variable that can take the values ± 1. Microscopic interactions are between nearest neighbors only, and the interaction favors the alignment of ths spins, that is, two neighboring spins with the same orientation contribute a negative energy. Competition between energy and entropy yields a phase transition at some critical temperature T_c, and for temperatures below T_c, a spontaneous magnetization appears. If a large system is rapidly quenched from above to below T_c, two-phase patterns consisting of "positive" and "negative" domains spontaneously appear. This model has been used as a prototypical example for a system that exhibits domain coarsening, see Bray (1994).

On a mesoscopic level, this system can be decribed by a coarse-grained free energy functional of the Ginzburg-Landau form,

$$F = \int_V \frac{1}{2} K \left(\nabla m\right)^2 + f(m) - hm, \tag{1}$$

where the continuous scalar field $m(\mathbf{x}, t)$ is the local (suitable nondimensionalized) magnetization, obtained by averaging over "coarse-graining" cells of a fixed size ℓ, K is a constant, $f(m)$ is the local free energy density, h is the (dimensionless) external magnetic field (a constant), and the integration is over the entire volume V of the system. The local free energy density $f(m)$ represents the free energy calculated by averaging over all microscopic configurations in a coarse-graining cell for fixed magnetization m. Therefore, its form obviously depends on the size of the coarse-graining cell. From simple mean-field arguments one obtains that an appropriate form of $f(m)$ is

$$f(m) = \frac{1}{2}a(T - T_c)m^2 + \frac{1}{4}bm^4, \tag{2}$$

where a and b are positive constants, T is the (uniform) temperature, and T_c is the critical temperature of the magnetic phase transition. Note that only even powers of m appear in $f(m)$ because of the up-down symmetry of the magnetic system. For $T < T_c$, $f(m)$ has a double-well structure, and each of the minima correponds to a phase with a spontaneous magnetization. Finally, the gradient square term represents the energy penalty due to magnetization inhomogeneities. The constant K is related to the interaction energy between spins, see for example Langer (1991).

As already mentioned, it is impossible in practice to carry out the coarse-graining procedure; for a recent approximation performed with the help of numerical calculations, see Bronchart et al. (2008). Nevertheless, from a phenomenological point of view, the above functional is reasonable, and the coefficients can be readily adjusted to match a physical system, with the help of four quantities that can in principle be measured: the magnetization, the susceptibility, the magnetic domain wall energy, and the domain wall mobility.

The equilibrium magnetization m^* simply corresponds to the value of m which minimizes $f(m)$ for zero magnetic field ($h = 0$); for the quartic potential given by Eq. (2), it is

$$m^* = \pm\sqrt{\frac{a(T_c - T)}{b}}. \tag{3}$$

Obviously, this minimum is shifted when a magnetic field is applied, since the magnetization tends to align with the external field. The exact value of $m^*(h)$ is the solution of a cubic equation. For small fields, the equation can be linearized around the zero-field solution $m^*(0)$ given by Eq. (3), which yields

$$m^*(h) \approx m^*(0) + \frac{h}{2a(T_c - T)} = m^*(0) + \frac{h}{f''(m^*(0))}, \tag{4}$$

where $f''(m)$ denotes the second derivative of $f(m)$ with respect to m. The magnetic susceptibility $\chi(T)$ is

$$\chi(T) = \left.\frac{dm^*(h)}{dh}\right|_{h=0} = \frac{1}{2a(T_c - T)} = \frac{1}{f''(m^*(0))}. \tag{5}$$

Obviously, the susceptibility diverges when $T \to T_c$, which is one of the signatures of a second-order phase transition.

In order to obtain the magnetic domain wall energy, we first have to determine the profile of the magnetization through a boundary between domains of positive and negative magnetizations. This profile is the minimum of the free energy functional, which can be obtained by evaluating

$$\frac{\delta F}{\delta m} = 0, \tag{6}$$

where the symbol δ denotes the functional derivative. The result of this operation is the equation

$$-K\partial_{xx}m + f'(m) = 0 \tag{7}$$

for a domain wall normal to the x direction, where $f'(m) = df/dm$. This equation has to be solved subject to the boundary condition $m \to m^*(0)$ for $x \to \infty$ and $m \to -m^*(0)$ for $x \to -\infty$. For the quartic potential given by Eq. (2), an explicit solution is available, which is

$$m_0(x) = -m^*(0)\tanh\left(\frac{x - x_0}{\xi}\right), \tag{8}$$

which describes a domain wall centered at the arbitrary position x_0, with

$$\xi = \sqrt{\frac{2K}{a(T_c - T)}} \tag{9}$$

being the characteristic thickness of the domain wall. It can be identified with the correlation length of the magnetization. Note that ξ diverges as $T \to T_c$, as expected for a correlation length in a second-order phase transition.

The quartic potential is widely used in the literature because of the existence of this simple analytic front solution. However, it should be stressed that any double-well potential would produce an equilibrium solution that has a similar structure: a front region in which the variation of m is rapid, surrounded by tails in which m approaches exponentially the minima of the potential.

The fundamental definition of the domain wall energy σ is that σ is the excess free energy due to the presence of a wall. In the absence of an external field h, the two homogeneous solutions $\pm m^*(0)$ have the same free energy. Then, σ can be calculated by the formula

$$\sigma = \frac{F(\text{interface}) - F(\text{homogeneous})}{S}, \tag{10}$$

where F(interface) and F(homogeneous) are the free energies with and without a domain wall, respectively, and S is the surface area along the wall. For a wall normal to the x direction as given by Eq. (8), S cancels out with the integrations along the y and z directions in the free energy functional, such that the above formula simplifies to

$$\sigma = \int_{-\infty}^{\infty} \frac{1}{2} K \left(\partial_x m_0\right)^2 + f(m_0(x)) - f(m^*)\, dx, \tag{11}$$

where we have used that the integrand of Eq. (1) simplifies to $f(m^*)$ for a homogeneous state.

The evaluation of this integral can be greatly simplified by the use of an identity that can be obtained from the equation of the equilibrium profile. Multiplication of both sides of Eq. (7) with $\partial_x m_0$ and integration of the result from $-\infty$ to a certain position x yields (after application of the chain rule)

$$\left[-\frac{K}{2} \left(\partial_x m_0\right)^2\right]_{-\infty}^{x} + \left[f(m_0)\right]_{-\infty}^{x} = 0. \tag{12}$$

Taking into account that $\partial_x m_0$ tends to zero for $x \to \pm\infty$ and that $m_0 \to \pm m^*$ for $x \to \pm\infty$, this simplifies to

$$\frac{K}{2} \left(\partial_x m_0\right)^2 = f(m_0) - f(m^*). \tag{13}$$

This relation is often called "equipartition relation" because it establishes that the two contributions to the free energy excess in Eq. (11) (gradient energy and potential energy) are of equal magnitude. Equation (13) can be used to obtain two important expressions for the surface free energy. In the first, the potential is eliminated, which yields

$$\sigma = \int_{-\infty}^{\infty} K \left(\partial_x m_0\right)^2\, dx. \tag{14}$$

This expression can be found in many textbooks. The evaluation of this integral requires the knowledge of the equilibrium profile. To circumvent this difficulty, Eq. (13) can also be used to eliminate the gradient energy from Eq. (11), which yields

$$\sigma = \int_{-\infty}^{\infty} 2[f(m) - f(m^*)]\, dx. \tag{15}$$

Furthermore, a simple transformation of Eq. (13) yields

$$\partial_x m_0 = \sqrt{\frac{2[f(m) - f(m^*)]}{K}}. \tag{16}$$

With the help of this relation, it is possible to change the variable of integration from x to m in Eq. (15) and to obtain

$$\sigma = \sqrt{K} \int_{-m^*}^{m^*} \sqrt{2[f(m) - f(m^*)]}\, dm. \tag{17}$$

Therefore, the surface free energy can be calculated from the double-well potential alone, without explicit knowledge of the equilibrium front profile. For the quartic potential of Eq. (2), its value is

$$\sigma = \frac{2\sqrt{2}}{3}\sqrt{K}(m^*)^2\sqrt{a(T_c - T)} = \frac{2\sqrt{2}}{3}\sqrt{K}\frac{[a(T_c - T)]^{3/2}}{b}, \tag{18}$$

where the second identity is obtained with the help of Eq. (3).

Before discussing the front mobility, it is necessary to specify the equation of motion for the magnetization m. In the Ising model, each spin can flip without any constraint. Hence, the local magnetization is a non-conserved quantity. According to non-equilibrium thermodynamics, its rate of change should therefore be proportional to the thermodynamic driving force,

$$\partial_t m = -\Gamma \frac{\delta F}{\delta m}. \tag{19}$$

In other words, the magnetization evolves such as to tend to its local free energy minimum. Here, Γ is a kinetic coefficient which will be taken constant (independent of m and T) for simplicity.

The equilibrium front given by Eq. (8) is stationary. This is natural since the double-well potential is symmetric, and hence the free energy densities of the two phases are identical. Obviously, in the presence of an external magnetic field, this is no longer the case: the phase in which the magnetization is aligned with the external field has a lower free energy and will hence grow at the expense of the other phase by a displacement of the domain wall. For small external fields, the free energy difference of the two phases is simply $\Delta f = 2m^*|h|$. Indeed, the equilibrium values of the magnetization are shifted by the magnetic field according to Eq. (4), but since the free energy curve is symmetric and hence its second derivative is the same in both wells, the shift is the same for both phases and the difference between the equilibrium magnetizations of the two phases remains the same as for zero field (to first order in h). It is to be expected that the velocity of the domain wall is proportional to Δf, the constant of proportionality being the front mobility.

The equation of motion for a steady-state planar front is readily written down in the frame attached to the front, in which $\partial_t = -v\partial_x$:

$$-v\partial_x m = -\Gamma\left[-K\partial_{xx} m + f'(m) - h\right]. \tag{20}$$

For a front which has negative magnetization for $x \to -\infty$, a positive velocity v (corresponding to a migration of the front towards positive x) is obtained for negative magnetic field ($h < 0$) since this favors the growth of the negative domain.

Equation (20) is a nonlinear ordinary differential equation for m. For a given magnetic field h, a solution exists only for specific values of v (for a general discussion of mathematical aspects of front solutions, see van Saarloos (2003)), and a solvability condition then determines v as a function of h. A general analytic solution of this problem is not available for the quartic potential. However, a good approximation can be obtained by the following procedure. Multiply both sides of Eq. (20) with $\partial_x m$ and integrate from $-\infty$ to ∞. The result is

$$v \int_{-\infty}^{\infty} (\partial_x m)^2 = \Gamma \left[-\frac{K}{2} (\partial_x m)^2 + f(m) - hm \right]_{-\infty}^{\infty}. \tag{21}$$

The evaluation of this integral requires the knowledge of the solution $m(x)$ of Eq. (20). However, for small fields it can be supposed that this profile will be close to the equilibrium one (a more detailed justification of this hypothesis will be given below), and the integrals can be evaluated with $m_0(x)$ instead of $m(x)$. With this approximation, since $\partial_x m_0 \to 0$ for $x \to \pm\infty$ and $f(m^*) = f(-m^*)$, the only non-vanishing term on the right hand side arises from the magnetic field, and is equal to $-2m^*h$, the free energy difference. The integral on the left hand side is equal to σ/K according to Eq. (14). Therefore,

$$v\frac{\sigma}{K} = -2\Gamma m^* h = \Gamma \Delta f \tag{22}$$

(recall that h is negative), and the front mobility is simply given by

$$M_{\text{front}} = \frac{\Gamma K}{\sigma}. \tag{23}$$

To summarize, the model has four unknown parameters: the coefficients a and b of the free energy function, the gradient energy coefficient K, and the kinetic constant Γ. They can be uniquely fixed using Eqs. (3), (5), (17), and (23) if the values of four physical quantities are known: the spontaneous equilibrium magnetization m^*, its dependence on the external field (the magnetic susceptibility), the magnetic domain wall energy σ and the front mobility. Note that, since Eq. (2) is a mean-field expression, it does not describe well the free energy density of any real magnetic system, especially close to the critical point. Therefore, the values of a and b that are obtained following the above procedure will certainly depend on the temperature.

Nevertheless, for any fixed temperature this model gives a very satisfactory description of the static and kinetic properties of magnetic domain walls.

It should be noted that the thickness of the domain walls, ξ, is not an independent parameter in this approach, but is given by Eq. (9). Furthermore, the capillary effect is automatically "built in" in this model. This can be easily seen if the above analysis is repeated for a spherical domain of negative magnetization surrounded by an infinite positive domain. Then, in spherical coordinates, the equation for a moving front becomes

$$-v\partial_r m = -\Gamma\left[-K\left(\partial_{rr}m + \frac{d-1}{r}\partial_r m\right) + f'(m) - h\right], \tag{24}$$

where d is the dimension of space. For large domains (of radius $R \gg \xi$), $1/r$ in the second term of the Laplacian can be replaced by $1/R$ since $\partial_r m$ is appreciable only in the interface, which is centered at $r = R$. Then, following the same steps as above yields

$$v\frac{\sigma}{\Gamma K} = -\frac{(d-1)\sigma}{R} - 2m^*h = -\frac{(d-1)\sigma}{R} + \Delta f. \tag{25}$$

For vanishing magnetic field, this identity becomes

$$v = -\Gamma K\frac{(d-1)}{R} = -M_{\text{front}}\sigma\frac{(d-1)}{R}, \tag{26}$$

which describes the so-called *motion by curvature*: the domain wall moves with a velocity that is proportional to its curvature $(d-1)/R$. It can be shown that this equation of motion remains valid even for non-spherical domains; $(d-1)/R$ then has to be replaced by the local mean curvature. Furthermore, local equilibrium ($v = 0$) is achieved if the magnetic field and the curvature are related by

$$2m^*h = -\frac{(d-1)\sigma}{R}. \tag{27}$$

This is called the capillary effect: the equilibrium value of an intensive quantity (here, the magnetic field) is shifted by an amount proportional to the curvature and to the surface free energy.

2.2 Binary lattice gas

Lattice gas models are a convenient setting to describe the thermodynamics and dynamics of two-component systems - be it fluids, polymers or metallic alloys. The systems for which they appear to be the most natural description are binary solid solutions, where the atoms of two different species occupy the sites of a common crystalline lattice. In the absence of vacancies, a site is occupied either by an A or a B atom. For the simplest case, in which atoms interact only when they occupy nearest neighbor sites, the Hamiltonian of the lattice gas model can be mapped onto the one of the Ising model discussed previously. The energetics (and hence the free energy functional and the thermodynamics that it implies) is hence completely equivalent to the one discussed above. In contrast, the dynamics is completely different: whereas, in the magnetic system, a spin can flip without any constraint, the numbers of A and B atoms are locally and globally conserved quantities, and an A atom cannot "flip" to become a B atom – it can only exchange places with a neighboring B atom.

The quantity that is analogous to the magnetization of the previous subsection is the concentration of one of the species – say, the B atoms. The fraction of B atoms within a coarse-graining cell is denoted by c ($0 \leq c \leq 1$). If the interaction between A and B atoms is weaker than the one between like atoms, *phase separation* will occur below a critical temperature T_c: the thermodynamic equilibrium state is a mixture of two phases, one rich in A and the other rich in B. In anamogy with the Ising model, we can write a free energy functional

$$F = \int_V \frac{K}{2} (\nabla c)^2 + f(c), \tag{28}$$

with

$$f(c) = \frac{1}{2} a(T - T_c)(c - c_c)^2 + \frac{b}{4}(c - c_c)^4 . \tag{29}$$

Here, c_c is the critical concentration at which phase separation first occurs; for a mixture with symmetric interaction constants between the two species, $c_c = 1/2$. Note that there is no term analogous to the external field term in the magnetic free energy functional. The quantity that is analogous to the magnetic field is the conjugate variable of the concentration, which is the chemical potential

$$\mu = \frac{\delta F}{\delta c}. \tag{30}$$

However, contrary to an external magnetic field that can be readily imposed, the application on an "external chemical potential" (for example, by establishing a contact with a "particle reservoir" of fixed properties) is difficult – what is normally imposed is the global concentration. The system then settles down to the thermodynamic equilibrium state, which is characterized by a homogeneous chemical potential.

As already mentioned above, the concentration is a locally conserved quantity. Therefore, is satisfies a continuity equation,

$$\partial_t c = -\nabla \cdot j, \tag{31}$$

where the mass current $\vec{j}$ is driven by the gradient of the chemical potential,

$$j = -M\nabla\mu, \tag{32}$$

with M the atomic mobility (here taken constant for simplicity). Combining these equations and the expression of the free energy yields

$$\partial_t c = M\nabla^2 \frac{\delta F}{\delta c} = M\nabla^2 \left[-K\nabla^2 c + f'(c)\right]. \tag{33}$$

This is called the Cahn-Hilliard equation, after Cahn and Hilliard (1958).

The profile of an equilibrium interface is now determined by the condition of constant chemical potential,

$$\mu = -K\nabla^2 c + f'(c) = \mu_{\rm eq}, \tag{34}$$

where the value of the equilibrium chemical potential $\mu_{\rm eq}$ can depend on the geometry. Bulk thermodynamics tells us that two homogeneous phases (of respective concentrations c_1 and c_2) are in equilibrium if two conditions are fulfilled: the chemical potentials must be equal, that is,

$$\mu_1 = f'(c_1) \equiv \mu_2 = f'(c_2), \tag{35}$$

and the grand potentials $\omega = f - \mu c$ must also be equal,

$$\omega_1 = f(c_1) - \mu_1 c_1 \equiv \omega_2 = f(c_2) - \mu_2 c_2. \tag{36}$$

When expressed graphically, these two equations describe a *common tangent*, a straight line that is tangent to the free energy curve in the two points c_1 and c_2. It is obvious that for the symmetric free energy function given by Eq. (29), the common tangent to the two wells has zero slope ($\mu_{\rm eq} = 0$). In that case, Eq. (34) becomes identical to Eq. (7). Furthermore, the equilibrium values of the composition simply correspond to the minima of f and are therefore given by the equivalent of Eq. (3):

$$c_1^* = c_c - \sqrt{\frac{a(T_c - T)}{b}} \qquad c_2^* = c_c + \sqrt{\frac{a(T_c - T)}{b}}. \tag{37}$$

However, this is not the only possible equilibrium state. Indeed, consider a spherical domain of the A-rich phase surrounded by an infinite domain of B-rich phase. Contrary to the spherical magnetic domain treated above, which shrinks according to Eq. (26) unless a magnetic field is applied, the inclusion of A-rich phase cannot evolve since the total number of A atoms (and hence the size of the inclusion) is conserved. The analogy with the magnetic system can nevertheless be exploited: the situation is equivalent to the case where the effects of curvature and magnetic field exactly compensate and the interface is stationary. Indeed, after subtraction of the constant $\mu_{\rm eq}$ on both sides of Eq. (34) and switch to spherical coordinates, it becomes formally identical to Eq. (24) with $v = 0$, with $\mu_{\rm eq}$ being equivalent to h. Hence, the equilibrium chemical potential is given in analogy to Eq. (27) by

$$(c_2 - c_1)\mu_{\rm eq} = -\frac{(d-1)\sigma}{R}, \tag{38}$$

where σ is now the interface free energy, computed as before using the analog of Eq. (17). It should be noted that the *sign* is important here: the chemical potential is *negative* for a spherical inclusion of A-rich phase, whereas it is *positive* for a spherical inclusion of B-rich phase in an A-rich matrix. Formally, this arises from the fact that the integration of c through the interface yields $c_1 - c_2$ on the left-hand side for a B-rich inclusion. Physically, this can be rationalized in the following way. The variable μ is the chemical potential of the B atoms. In the bulk phases (far away from any interface), where the concentration is homogeneous (or at least slowly varying), μ is a function of the local concentration only (the Laplacian in Eq. (30) is negligibly small). Since $f(c)$ is a convex function in the neighborhood of the equilibrium solutions, μ is a monotonously increasing function of c. Therefore, $\mu > 0$ for $c > c_i$ in both phases ($i = 1, 2$). But phase separation takes place because the interactions between like atoms are stronger that between distinct atoms. Since a B atom on the surface of a spherical B-rich inclusion is on average linked by fewer bonds to other B atoms than on a planar interface, its departure into the matrix phase is facilitated, which results in a supersaturation of the matrix phase, that is, $c > c_1$. Of course, the reverse is true for an inclusion of A-rich phase.

The two cases can be easily unified by defining a curvature κ that can be positive and negative. This can be easily done in the following way. In the interfaces, the unit vector

$$\hat{n} = -\frac{\nabla c}{|\nabla c|} \tag{39}$$

is normal to the interfaces and points into the A-rich phase. According to differential geometry, the local curvature of the interface is given by

$$\kappa = \nabla \cdot \hat{n}, \tag{40}$$

which is positive if the B-rich domain is locally convex. With the help of this quantity,

$$\mu_{\rm eq} = \frac{(d-1)\sigma}{c_2 - c_1}\kappa. \tag{41}$$

To complete the picture, we still need the values of c_1 and c_2, which depend of $\mu_{\rm eq}$. For large domains ($R \gg \xi$), where $\mu_{\rm eq}$ is small, we can use a Taylor expansion around $\mu_{\rm eq} = 0$ (where $c_i = c_i^*$) and find

$$\mu_{\rm eq} = 0 + f''(c_i^*)(c_i - c_i^*). \tag{42}$$

For the symmetric double-well potential, the second derivatives are identical, and hence the concentration difference is independent of the curvature and equal to $c_2^* - c_1^*$ (up to first order in the shift). Thus, we find finally

$$c_i = c_i^* + \frac{(d-1)\sigma}{f''(c_i^*)(c_2^* - c_1^*)}\kappa, \tag{43}$$

the so-called Gibbs-Thomson law. Of course, Eq. (42) is completely analogous to Eq. (4), and a generalized susceptibility can be defined by $\chi = 1/f''(c_i^*)$.

The equation of motion for the interfaces is more complicated than for the magnetic model. Because of the conservation law, the motion of an interface is determined by the local flux of B atoms that arrive or depart from the interface. Therefore, no simple local law of motion can be written down, and the calculation of the interface mobility is more involved (see Elder et al. (2001) for a detailed discussion). Nevertheless, the above developments permit to understand a phenomenon of major importance that takes place in alloys, namely, coarsening (Ostwald ripening). Indeed, consider several B-rich inclusions of different sizes in a common A-rich matrix. Since, according to Eq. (42), the chemical potential of B atoms is higher on the surface of smaller inclusions (they have higher cuvature), atoms will diffuse from smaller to larger inclusions. The small inclusions will hence "evaporate" with time. In a finite system, the final equilibrium state is a single spherical inclusion. In an infinite system, the process continues forever and gives rise to a scaling behavior with time that has been much studied, see Bray (1994).

2.3 Second gradient theory for fluids

In the second gradient theory, the dynamics of fluids and of fluid-vapor interfaces is described in terms of the density field ρ, which directly acts as an order parameter to distinguish between the two phases. The model takes its name from the fact that it can be obtained from a free energy functional that depends not only on the local fluid density, but also on its gradient, in perfect analogy to the models discussed in the previous sections,

$$F = \int_V \frac{1}{2} K \left(\nabla\rho\right)^2 + f(\rho), \tag{44}$$

where $f(\rho)$ is the local free energy density and K is a constant.

Since this model is discussed in great detail for example in Jamet et al. (2001) and also elsewhere in this volume *Roberto: put cross-reference to your contribution if appropriate*, only a few salient facts that are important in the present context will be reviewed. In this case, the free energy function $f(\rho)$ does not have any obvious symmetries, such that on first sight it does not seem appropriate to approach it by a simple symmetric double-well potential. However, if there is liquid-vapor coexistence, the function $f(\rho)$ necessarily has concave parts. Indeed, a thermodynamic chemical potential can be defined by $\tilde{\mu} = \partial f/\partial\rho$, and the conditions for liquid-vapor coexistence are exactly equivalent to those for the binary alloy given above. Therefore, the coexistence densities are obtained by a common tangent construction to the free energy curve. The result of this construction are the coexistence densities as well as the equilibrium chemical potential $\tilde{\mu}_{\rm eq}$. Then, the change of variables $\tilde{\mu} \to \tilde{\mu} - \tilde{\mu}_{\rm eq}$ and $f(\rho) \to f(\rho) - \tilde{\mu}_{\rm eq}\rho$ will transform the free energy density into a double-well potential with minima of equal energy. If a fluid close to its critical point is considered, then this modified free energy density can again be approximated by the symmetric double well potential.

The equation of motion for the fluid density cannot directly be obtained by a variational derivative of F. Instead, ρ satisfies a set of balance equations for mass, momentum, and energy, that has to take into account inertia, viscous dissipation, and the mechanical effect of the surface tension, which creates a capillary force. The resulting equations can be found for example in Jamet et al. (2001). For our discussion, two important points are that the surface tension of this model is given by the analog of Eq. (14),

$$\sigma = \int_{-\infty}^{\infty} K \left(\partial_x \rho_0\right)^2 \, dx, \tag{45}$$

where $\rho_0(x)$ is the density profile across a planar interface normal to the x direction. Of course, this integral can be transformed into one that contains only the free energy density by the same steps as described for the magnetic model in Sec. 2.1. But the free energy function obviously also determines the equations of state of the fluid in the two phases, and thus various thermodynamic quantities such as the isothermal compressibility. In summary, both the bulk and the surface properties of the model are determined by the choice of the free energy function $f(\rho)$.

2.4 Model C for solidification

The equations of motion for the magnetic model and the alloy model presented above are named "model A" and "model B", respectively, in the review of Hohenberg and Halperin (1977) on dynamic critical phenomena. The idea behind this classification is that in the vicinity of a critical point, static and dynamic properties of many very different systems are universal, and universality classes can be distinguished based on the dimensionality and conservation laws of the involved variables (order parameters). The magnetization is a non-conserved scalar order parameter (model A), the composition is a conserved scalar one (model B). "Model C" in this classification is a model with two scalar fields, one of which is conserved, and the other is non-conserved. The first phase-field models for solidification put forward by Langer (1986), Fix (1983) or Collins and Levine (1985) (see Karma and Rappel (1998) for a brief historic overview of the development of the phase-field model) were just reformulations of this model, with the idea in mind that the non-conserved field corresponds to a structural order parameter (which distinguishes between liquid and solid), whereas the conserved variable represents the diffusion field that limits crystal growth (heat or chemical components).

A starting point for this model is the free energy functional

$$F = \int_V \frac{K}{2}(\nabla\phi)^2 + H f_{\mathrm{dw}}(\phi) + Xu\phi, \tag{46}$$

where ϕ is the phase field, f_{dw} is a dimensionless double-well function – in the following, we will use

$$f_{\mathrm{dw}}(\phi) = \frac{1}{4}(1-\phi^2)^2. \tag{47}$$

Furthermore, K, H, and X are constants, and u is a dimensionless diffusion field (to be defined more precisely below). Since the phase field represents a non-conserved structural order parameter, it obeys the model A dynamics,

$$\partial_t\phi = -\Gamma\frac{\delta F}{\delta\phi}, \tag{48}$$

with the kinetic rate constant Γ. The field u obeys a diffusion equation with a source term,

$$\partial_t u = D\nabla^2 u + \frac{1}{2}\partial_t\phi, \tag{49}$$

where D is a diffusion coefficient.

For the solidification of a pure substance, $u = C(T - T_m)/L$, where T is the local temperature, T_m the melting temperature, and C and L are the specific heat and the latent heat per unit volume, respectively. With this interpretation in mind, it is easy to understand the motivation of the above equations. When $T = T_m$, $u = 0$, and the free energy density of the order parameter is described by a symmetric double-well potential. As a consequence, there is no driving force for solidification since the two phases have the same free energy density. In contrast, for $u \neq 0$, the coupling term $X\phi u$ breaks the symmetry and introduces a free energy difference between the phases. The difference in the free energy density is $\Delta f = 2Xu$ (since the two equilibrium values of ϕ are $\phi^* \approx \pm 1$ and thus their difference is two). Equating this to the free energy difference obtained from standard bulk thermodynamics, $\Delta f = L(T - T_m)/T_m$, and using the definition of u, we can identify

$$X = \frac{L^2}{2CT_m}. \tag{50}$$

This free energy difference induces a motion of the interfaces, since the phase with the lower free energy will form at the expense of the other one. Thus, liquid is transformed into solid (or the reverse), and the latent heat of melting is released (or consumed). This fact is described by the source term in Eq. (49).

A comparison of Eq. (46) to Eq. (1) shows that the free energy functional of model C is identical to the one of the magnetic model, with the free energy density $f(m)$ replaced by $Hf_{\rm dw}(\phi)$, and the magnetic field h replaced by $-Xu = -Lu/2$. From this, we can immediately deduce that the equilibrium values of ϕ actually depend on the temperature u,

$$\phi^*(u) \approx \pm 1 - \frac{uX}{Hf''_{\rm dw}(\pm 1)}. \tag{51}$$

Therefore, ϕ *has a non-trivial behavior in the bulk.* As discussed for example by Penrose and Fife (1990), the phase field can indeed be interpreted as a physical quantity (entropy or energy density), such that this model is actually a diffuse-interface model after the classification introduced in Sec. 1. As will be discussed in more detail below, this model is too simple for accurate simulations of crystal growth. But it can still serve as an excellent starting point for the introduction of a more sophisticated model.

3 Solidification of a pure substance

In this section, I will introduce and discuss in some detail the problem of solidification of a pure substance, which has served as one of the standard testing grounds for the development of phase-field modelling. I will start by stating the classic free boundary problem of solidification, and then give the standard phase-field formulation. The differences to the diffuse-interface models described in the previous section will be discussed in some detail. Then, I will describe the method of matched asymptotic expansions that can be used to relate the sharp-interface and phase-field formulations of the solidification problem, and review some salient results obtained with this methodology.

3.1 Sharp-interface formulation

Consider the solidification of a pure substance from its melt. Since the relative change of density upon solidification is small, the density will be assumed to be constant and equal for solid and liquid. Then, no mass transport is needed for crystal growth, which is therefore limited by heat transport, since the latent heat of crystallization that is released upon freezing needs to be evacuated from the growing crystal. Heat transport is assumed to take place by diffusion only; natural convection is neglected. This is a big simplification, which is strictly speaking valid only under microgravity conditions. It is justified by several independent reasons: (i) the purely diffusion-limited case is one of the rare examples where analytical solutions and theories have been developed, and can therefore be studied in its own right, (ii) data from experiments in microgravity are available, see Glicksman et al. (1994), and (iii) this is not a fundamental limitation of the phase-field approach. Indeed, models by Beckermann et al. (1999) or Anderson et al. (2000) that incorporate convection are available, but require much larger computational resources. For the pedagogic exposition intended here, it is thus sufficient to consider the purely diffusive case.

Currents of heat are driven in the liquid and the solid by temperature gradients,

$$j_Q = -k_\nu \nabla T = -C_\nu D_\nu \nabla T, \tag{52}$$

where $\nu = l, s$ labels the two phases, and k_ν is the heat conductivity of phase ν, which is written in the second identity as the product of the specific heat per unit volume C_ν and the thermal diffusion coefficient D_ν. Energy conservation implies that the internal energy density e satisfies

$$\partial_t e = -\nabla \cdot j_Q. \tag{53}$$

We can exploit the definition of the specific heat per unit volume, $C_\nu = de/dT$, to transform this equation for e into one for T. For constant specific heats and constant diffusion coefficients, we obtain simple diffusion equations in both phases,

$$\partial_t T = D_\nu \nabla^2 T. \tag{54}$$

At the interfaces, heat conservation gives rise to a Stefan boundary condition,

$$L v_n = \hat{n} \cdot [j_Q] = \hat{n} \cdot \left[C_s \left. \nabla T \right|_s - C_l \left. \nabla T \right|_l \right], \tag{55}$$

where L is the latent heat of melting per unit volume, v_n is the normal velocity of the interface (counted positive for a growing solid), $\hat{n}$ is the normal vector to the interface pointing into the liquid, and $[j_Q]$ is the difference in heat currents between the two sides of the interface, which is explicitly given in the second identity. The left hand side of this equation represents the heat generated (or absorbed) by a moving interface; this heat is transported by the heat currents in the adjacent material, which are given by the right hand side.

The system of equations is completed with the specification of a boundary condition for the temperature at the interface,

$$T_{\text{int}} = T_m - \frac{\sigma T_m}{L} \kappa - \frac{v_n}{\mu_k}. \tag{56}$$

Here, T_m is the melting temperature, σ the interface free energy, κ the interface curvature, and μ_k the interface mobility. The interface temperature deviates from the bulk thermodynamic equilibrium value T_m by two terms. The first is the capillary shift of the melting temperature (Gibbs-Thomson effect). The second is due to interface kinetics. The fact that μ_k is indeed a mobility can be seen by inverting this equation to obtain v_n, which yields

$$v_n = \mu_k \left[T_{\text{int}} - \left(T_m - \frac{\sigma T_m}{L} \kappa \right) \right]. \tag{57}$$

The interface velocity is hence proportional to the difference of the interface temperature and the curvature-dependent local equilibrium interface temperature, and the proportionality factor is thus a mobility.

It should be noted that the above generalized Gibbs-Thomson law is *isotropic*. In general, the interfacial properties (surface free energy and interface mobility) of a crystal-melt interface are anisotropic (that is, their values depend on the orientation of the surface with respect to the crystallographic axes). While the magnitude of this anisotropy is small for microscopically rough interfaces (as usually found in metals), its presence is crucial for the selection of dendrite growth directions. Nevertheless, for most of the following we will consider isotropic interfaces for simplicity.

For further simplification, we will consider first the *symmetric model*, in which the diffusion coefficients and the specific heats of solid and liquid are assumed to be identical:

$$C_s = C_l \equiv C \qquad D_s = D_l \equiv D. \tag{58}$$

Then, the above set of equations can be simplified and rewritten in terms of the dimensionless temperature field

$$u = \frac{T - T_m}{L/C}. \tag{59}$$

The result is

$$\partial_t u = D\nabla^2 u \qquad \text{in the bulk,} \tag{60}$$

$$v_n = D\hat{n} \cdot \left[\nabla u|_s - \nabla u|_l \right] \qquad \text{at the interface} \tag{61}$$

$$u_{\text{int}} = -d_0\kappa - \beta v_n, \tag{62}$$

where the capillary length d_0 is given by

$$d_0 = \frac{\sigma T_m C}{L^2}, \tag{63}$$

and β is a kinetic coefficient with dimension of inverse velocity,

$$\beta = \frac{C}{\mu_k L}. \tag{64}$$

3.2 Phase-field model

There are several ways to motivate the construction of a suitable phase-field model for this problem. One is to start from bulk thermodynamics. The idea is to introduce a phase field that distinguishes between the two possible phases (solid and liquid). The two values of the phase field that represent solid and liquid can be arbitrarily chosen – we will adopt $\phi = 1$ for the solid and $\phi = -1$ for the liquid. Next, a free energy density needs to be constructed which combines the free energies of the pure phases. For the present case, we may start from the internal energy densities e. Since we have assumed the specific heat $C = de/dT$ to be constant and equal for the two phases, the two curves $e(T)$ are just two parallel straight lines. Furthermore, at the melting temperature T_m, the two free energy densities are equal,

$$f_s = e_s - T_m s_s = f_l = e_l - T_m s_l, \tag{65}$$

with s_s and s_l the entropy densities of the two phases. Furthermore, the definition of the latent heat of melting yields

$$L = e_l(T_m) - e_s(T_m) = T_m(s_l - s_s). \tag{66}$$

Then, up to a constant (which can be omitted) the internal energy density as a function of T and ϕ is

$$e(T,\phi) = C(T - T_m) - \frac{g(\phi)}{2}L, \tag{67}$$

where $g(\phi)$ is an interpolation function that satisfies $g(\pm 1) = \pm 1$. Dividing both sides of this equation by L, we can express the dimensionless internal energy density $\tilde{e} = e/L$ in terms of the dimensionless temperature u introduced above,

$$\tilde{e} = u - \frac{g(\phi)}{2}. \tag{68}$$

The equations of motion of the phase-field model can now be obtained by a variational procedure in the variables $\tilde{e}$ and ϕ from the free energy functional

$$F = \int_V \frac{1}{2}K\left(\nabla\phi\right)^2 + Hf_{\mathrm{dw}}(\phi) + \frac{X}{2}\left(\tilde{e} + \frac{g(\phi)}{2}\right)^2 \tag{69}$$

where X is again given by Eq. (50). Indeed, energy is a conserved quantity and hence the internal energy density obeys model B dynamics,

$$\partial_t \tilde{e} = M\nabla^2 \frac{\delta F}{\delta \tilde{e}}, \tag{70}$$

where M is a suitable mobility. In contrast, the phase-field ϕ is a non-conserved variable and thus obeys model A dynamics,

$$\partial_t \phi = -\Gamma \frac{\delta F}{\delta \phi}. \tag{71}$$

Carrying out the functional derivatives, rearranging terms, and using the various definitions introduced previously yields the two equations

$$\partial_t u = D\nabla^2 u + \frac{1}{2}\partial_t g(\phi), \tag{72}$$

$$\Gamma^{-1}\partial_t \phi = K\nabla^2\phi - Hf'_{\mathrm{dw}}(\phi) - \frac{X}{2}g'(\phi)u. \tag{73}$$

For $g(\phi) = \phi$, this is identical to the simple model C presented in Sec. 2.4.

In the following, we will also discuss a slightly more general model, in which Eq. (72) is replaced by

$$\partial_t u = D\nabla^2 u + \frac{1}{2}\partial_t h(\phi), \tag{74}$$

where $h(\phi)$ is a function that satisfies $h(\pm 1) = \pm 1$. This function describes how the latent heat is released through the moving interface and will therefore be called "heat source function". Of course, as soon as $h(\phi) \neq g(\phi)$, the model cannot any longer be obtained by a variational procedure from a free energy functional. However, the equation of motion for ϕ can still be obtained from the functional

$$F' = \int_V \frac{1}{2}K\,(\nabla\phi)^2 + Hf_{\rm dw}(\phi) + \frac{X}{2}ug(\phi). \tag{75}$$

In Karma and Rappel (1998), the case $h(\phi) = g(\phi)$ is called the *variational formulation*, whereas the more general case $h(\phi) \neq g(\phi)$ is called the *isothermal variational formulation*.

3.3 Discussion: phase-field vs. diffuse-interface models

The only difference between model C and the phase-field model presented above is the presence of the interpolation function $g(\phi)$ introduced in Eq. (67). The choice of this function has extremely important consequences which merit to be discussed in more detail.

To begin, let us analyze a little further the simple model C. It actually has a major default, which is a consequence of Eq. (51). Indeed, since according to this equation, the equilibrium value of ϕ depends on the temperature, a change of temperature induces a change of the phase field even deep inside the bulk (far away from any interfaces). But this generates sources of latent heat in the bulk according to Eq. (49). This is clearly physically incorrect: latent heat should only be generated or absorbed at moving interfaces according to Eq. (55) of the sharp-interface formulation.

The interpolation function $g(\phi)$ can be used to cure this problem. Repeating the calculations leading to Eq. (51) for the above phase-field model, we find

$$\phi^*(u) \approx \pm 1 - \frac{Xg'(\pm 1)u}{2Hf''_{\rm dw}(m^*(0))}. \tag{76}$$

This, the equilibrium values can be kept *exactly equal* to ± 1 if $g(\phi)$ is chosen such as to have vanishing derivatives at the values of ϕ that correspond to the minima of the double-well potential. For a graphic visualization, the sum $Hf_{\rm dw}(\phi) + (X/2)g(\phi)u$ is plotted in Fig. 1 for suitable values of H,

L and u. The function g actually describes how the double-well potential is *tilted* for $T \neq T_m$. The double-well potential has to be tilted such as to keep the positions of the minima at $\pm$ 1 for arbitrary values of u. For this, the simple linear function $g(\phi) = \phi$ used in model C is not suitable. With the additional requirement that $g(\pm 1) = \pm 1$, the simplest choice is the third-order polynomial

$$g(\phi) = \frac{3}{2}(\phi - \phi^3/3). \tag{77}$$

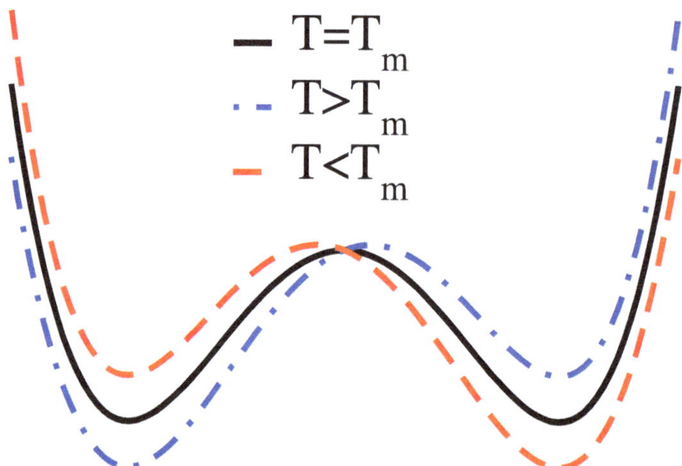

Figure 1. Plot of the tilted double-well potential, with the tilting function $g(\phi)$ given by Eq. (78), for three values of the temperature. The tilt is such that the positions of the minima do not change.

The introduction of the tilting function $g(\phi)$ seems to be only a small change, but it actually has big consequences for the interpretation of the phase field. Since the equilibrium values of ϕ are $\phi^* = \pm 1$ independently of the temperature, it is impossible to identify the phase field with any physically measurable quantity. Indeed, let us first examine the thermodynamic variables. The phase field exhibits a jump through the interface, just like an extensive quantity. For a pure substance under the assumption of constant density, there are just two such quantities: internal energy density and entropy density. However, the derivatives of both of these quantities with respect to temperature are well-defined quantities: $de/dT = C$, the specific heat per unit volume, as already stated above, and $ds/dT = C/T$ (using the chain rule, $ds/dT = (ds/de)(de/dT)$, and the thermodynamic definition $ds/de = 1/T$). Since $d\phi^*/dT = 0$, the phase field cannot be identified with either of these quantities. Furthermore, the original introduction of the phase field was motivated by the idea of a structural order parameter that distinguishes between solid and liquid. Such an order parameter is either introduced phenomenologically in the spirit of Landau theory, or defined by some microscropic prescription, for example in molecular dynamics simulations, see Hoyt et al. (2003). However, in both cases the equilibrium values of such a quantity also depend on the temperature.

In conclusion, in this model the phase field can only be interpreted as a smoothed indicator field, without direct thermodynamical meaning. It will now be shown that for the purpose of quantitative modeling, this is actually an advantage because is allows to some extent to separate the adjustment of interface and bulk properties. For illustration, let us first calculate the equilibrium front profile. Since equilibrium of a planar front can only occur at $T = T_m$, we have $u = 0$ and the equation of the equilibrium phase field profile becomes

$$K\partial_{xx}\phi - Hf'_{\mathrm{dw}}(\phi) = 0. \tag{79}$$

Dividing both sides of this equation by H, posing

$$W^2 = \frac{K}{H} \tag{80}$$

and carrying out the derivative of f_{dw}, this becomes

$$W^2\partial_{xx} + \phi - \phi^3 = 0, \tag{81}$$

the explicit solution of which is

$$\phi(x) = \tanh\left(\frac{x}{\sqrt{2}W}\right). \tag{82}$$

Therefore, up to a prefactor, W is the front thickness. Repeating the same steps as for the diffuse-interface models, it is easy to obtain the surface tension

$$\sigma = \frac{2\sqrt{2}}{3}\sqrt{KH} = IWH, \tag{83}$$

where $I = 2\sqrt{2}/3$. Qualitatively, this result could have been obtained by a simple dimensional analysis. Indeed, the coefficients K and H in the functional of Eq. (75) have dimensions of energy per unit length and energy per unit volume, respectively. Therefore, the square root of their ratio has dimension of length, and the square root of their product has dimension of energy per unit surface, which is the dimension of a surface free energy. The numerical prefactor I comes from the integration of the double-well potential according to the analog of Eq. (17). Therefore, a different choice for the double-well function will change the interface profile and the value of I, but not the scaling of the physical quantities. The second identity of Eq. (83) has a particularly clear and simple interpretation: the surface free energy scales as the barrier height h in the free energy functional times the front thickness W. Indeed, the parts of the system located in the interface are energetically penalized by an amount of order H and contribute to the surface free energy excess over a thickness W.

Consider now a physical system that has interfaces of a certain physical thickness ξ and surface free energy σ. If, for numerical reasons, we wish to represent this system by a model that has *the same* surface free energy, but a *much larger* interface thickness, $W \gg \xi$, this is easy to achieve in the phase-field model: since σ and W are fixed by the coefficients K and H, it is sufficient to change both at the same time. For example, it is possible to keep σ constant while increasing W by increasing H but desceasing K. At the same time, this change does not affect the thermodynamic behavior of the bulk phases, which is entirely given by the third term in the functional of Eq. (75) that is not affected by changes in K or H.

In contrast, the same operation in model C leads to difficulties. The interfacial properties actually satisfy the same relations as discussed just above, since they do not depend on the presence of the interpolation function $g(\phi)$. But for this model, in addition, the bulk properties of the *physical quantity* that describes the front profile are also determined by the double-well potential according to Eq. (51). For a correct modelling, the variation of ϕ^* with temperature has to be identified with the corresponding thermodynamical quantity, which already fixes the constant H according to Eq. (51). Therefore, a simple rescaling of the double-well potential by the change of H is excluded. It is still possible to modify the surface free energy while keeping the bulk properties constant by more complicated modifications of the double-well potential (keeping the wells unchanged and changing the barrier height, for instance), as tempted for example by Jamet et al. (2001) for a simple diffuse-interface fluid, but the range of the possible modifications is limited.

3.4 Matched asymptotic analysis

Up to now, we have not discussed interface kinetics in the phase-field model. The reason is that its evaluation is considerably more complicated than in the simple models discussed so far. We were able to calculate a simple approximation for the interface mobility in the magnetic model of Sec. 2.1 because the driving force for interface motion was an externally imposed magnetic field, which was homogeneous and constant. In the phase-field model for solidification, the driving force for the phase change is the deviation of the local temperature from the melting temperature. But the temperature field obeys itself a differential equation, is non-homogeneous and time-dependent.

The standard procedure to understand the behavior of a phase-field model and to establish the relation between the model parameters and the physical parameters of the system to be modeled is the method of *matched asymptotic expansions*, a variant of the boundary-layer methods with which the reader might be familiar from fluid dynamics. The basis for these calculations is a *separation of scales* between the front thickness and the other characteristic scales of the problem. Indeed, the problem of solidification can successfully be described by a sharp-interface formulation, in which the interfaces are assimilated to mathematical surfaces without thickness, *because* the physical interfaces have an intrinsic thickness of a few interactomic distances, which is much smaller than any other scale present in the problem.

If this problem is represented by a phase-field model, it is immediately clear that space can be divided in two types of regions. In the bulk phases the phase field is equal to or close to ± 1, the equation of motion for the phase field is trivial, and the temperature field obeys a simple diffusion equation. The bulk phases are separated by a thin but finite layer of *interface*, in which the phase field obeys a non-trivial nonlinear equation that generates source terms in the diffusion equation for the temperature field. Whereas this layer has a complicated overall geometry, the *local* interface properties (curvature, interface velocity) vary only slowly along this layer if scale separation is realized. The central idea of the matched asymptotics is then to solve the non-trivial interface equations in the simple situation where the velocity and the interface curvature are locally constant, and to match this solution to the large-scale solution of the simple diffusion equation away from the interfaces. In this way, the nonlinear interface equations provide *boundary conditions* for the diffusion equation in the bulk.

Technically, the complete calculation is quite cumbersome. It has been presented in detail in several publications, see for example Karma and Rappel (1998), Almgren (1999), or Echebarria et al. (2004). Rather than repeating these lengthy developments here, I will restrict the presentation to a simpler problem: the motion of a planar interface with constant velocity. This calculation yields the interface mobility, and illustrates all the steps of the asymptotics. The main physical phenomenon that is missing fron this calculation is the capillary effect. However, its understanding poses no new challenge since it can be calculated at equilibrium, which implies that the temperature field is constant. Therefore, the calculation follows the same lines as for the magnetic model in Sec. 2.1.

It is convenient to make the evolution equations non-dimensional. Let us start by dividing the evolution equation for the phase field, Eq. (73), by the constant H. The result is,

$$\tau \partial_t \phi = W^2 \nabla^2 \phi - f'_{\rm dw}(\phi) - \lambda g'(\phi) u, \tag{84}$$

where we have defined the phase-field relaxation time $\tau = 1/(\Gamma H)$ and the dimensionless coupling constant

$$\lambda = \frac{X}{2H} = \frac{L^2}{2HCT_m}. \tag{85}$$

This is the standard form of the phase-field equation used in many publications.

Next, we need to choose a "macroscopic" scale for non-dimensionalization of the equations. We follow Almgren (1999) and choose the capillary length d_0, although a different choice yields a simpler calculation for the symmetric model according to Karma and Rappel (1998); however, it is easier to generalize the present calculation to unequal diffusivities. The corresponding time scale is then d_0^2/D. After switching to dimensionless variables, the equations for a planar interface that advances with a constant dimensionless velocity v in the positive x direction are

$$\alpha\epsilon^2 \partial_t \phi = -\alpha\epsilon^2 v \partial_x \phi = \epsilon^2 \nabla^2 \phi - f'_{\rm dw}(\phi) - \lambda u g'(\phi), \tag{86}$$

where $\alpha = D\tau/W^2$, and $\epsilon = W/d_0$ is the scaled interface thickness. The constant λ can actually be eliminated in favor of ϵ. To this end, we invert Eq. (83) to find $H = \sigma/(IW)$, and then eliminate σ by expressing it in terms of the capillary length, Eq. (63). As a result, we obtain

$$\lambda = \frac{I}{2}\frac{W}{d_0} = \frac{I}{2}\epsilon, \tag{87}$$

and the equation for the phase field becomes

$$-\alpha\epsilon^2 v \partial_x \phi = \epsilon^2 \nabla^2 \phi - f'_{\rm dw}(\phi) - \epsilon I u g'(\phi)/2. \tag{88}$$

The equation for the temperature field in the moving frame is simply

$$\partial_t u = -v\partial_x u = \nabla^2 u + \frac{1}{2}\partial_t h(\phi) = \nabla^2 u - \frac{v}{2}\partial_x h(\phi). \tag{89}$$

The dimensionless interface thickness is in fact the parameter that quantifies the scale separation, since it represents the ratio of the interface thickness W to the "external" physical scales. It is therefore chosen as an expansion parameter in two different expansions in the "outer" and "inner" region,

$$\phi = \hat{\phi}_0 + \epsilon\hat{\phi}_1 + \epsilon^2\hat{\phi}_2 + \dots \quad \text{outer region} \tag{90}$$

$$\phi = \phi_0 + \epsilon\phi_1 + \epsilon^2\phi_2 + \dots \quad \text{inner region} \tag{91}$$

$$u = \hat{u}_0 + \epsilon\hat{u}_1 + \epsilon^2\hat{u}_2 + \dots \quad \text{outer region} \tag{92}$$

$$u = u_0 + \epsilon u_1 + \epsilon^2 u_2 + \dots \quad \text{inner region.} \tag{93}$$

The outer expansion can be directly inserted in the above equations, and the result is then solved successively order by order. To order zero in ϵ, we obtain from Eq. (88)

$$0 = f'_{\mathrm{dw}}(\hat{\phi}_0), \tag{94}$$

which indicates (as anticipated) that the phase field should assume one of its bulk equilibrium values. Since, thus, $\hat{\phi}_0$ is independent of time, the source term on the right hand side of Eq. (89) is zero, and $\hat{u}_0$ obeys the simple diffusion equation. It is easy to show that these results extend to higher orders in the expansion.

The inner expansion is intended to describe the details of the interface on the scale of the interface thickness W itself. To "zoom in" on this region, we define a *stretched coordinate*

$$\eta = \frac{x}{\epsilon} \tag{95}$$

approprate to this scale. Since, upon the change of variable from x to η, each spatial derivative contributes a factor ϵ^{-1}, the formal effect of this operation is to change the powers of ϵ in each term of the equations. The resulting equations, valid in the inner region, are

$$-\epsilon\alpha v\partial_\eta\phi = \partial_{\eta\eta}\phi - f'_{\mathrm{dw}}(\phi) - \epsilon\frac{I}{2}ug'(\phi), \tag{96}$$

$$-\epsilon^{-1}v\partial_\eta u = \epsilon^{-2}\partial_{\eta\eta}u - \frac{1}{2}\epsilon^{-1}v\partial_\eta\phi. \tag{97}$$

Indeed, comparing Eq. (96) to Eq. (88), it is obvious that the order of the Laplacian term on the right hand side has been changed. This expresses the fact that the spatial derivatives of ϕ balance the derivative of the double-well potential on the scale of the interface thickness, which precisely leads to a smooth interface.

The explicit calculation of the inner expansion will be given below. Before proceeding, it is useful to state the matching conditions, which are formally defined in the limit when $\epsilon \to 0$,

$$\lim_{\eta\to\pm\infty} \left[u_0(\eta) - \hat{u}_0|^{\pm} \right] = 0, \tag{98}$$

$$\lim_{\eta\to\pm\infty} \left[u_1(\eta) - \hat{u}_1|^{\pm} - \eta\, \partial_x u_0|^{\pm} \right] = 0, \tag{99}$$

where $\hat{u}_n|^{\pm}$ are the limits of the outer fields when the interface is approached from the liquid (+) and the solid (−) side. The conditions for ϕ and for higher order terms of the expansions are similar. The meaning of these formulas is: the asymptotes of the inner fields, extrapolated to the outer region, must match to the local behavior of the outer fields when the interface is approached.

The explicit calculation proceeds as follows. Inserting the expansion for the phase field into Eq. (96) yields at order zero in ϵ

$$0 = \partial_{\eta\eta}\phi_0 - f'_{\rm dw}(\phi_0), \tag{100}$$

which is just the equation for the equilibrium phase-field profile and has the solution

$$\phi_0 = -\tanh\frac{\eta}{\sqrt{2}} \tag{101}$$

for a solid growing towards positive η. For the temperature field, we find at order ϵ^{-2}

$$0 = \epsilon^{-2}\partial_{\eta\eta}u_0. \tag{102}$$

Taking into account the matching conditions, which imply that $\partial_\eta u_0$ vanishes for $\eta \to \pm\infty$, we can conclude that u_0 is simply a constant, the value of which will be fixed below.

The phase-field equation at order ϵ reads

$$-\alpha v \partial_\eta\phi_0 = \partial_{\eta\eta}\phi_1 - f''_{\rm dw}(\phi_0)\phi_1 - \frac{I}{2}g'(\phi_0)u_0, \tag{103}$$

where $f''_{\rm dw}$ denotes the second derivative of the double-well function $f_{\rm dw}$. This is a differential equation for the function ϕ_1. It can be rewritten under the form

$$L_1\phi_1 = f_1(\eta), \tag{104}$$

with the linear operator $L_1 = \partial_{\eta\eta} - f''_{\rm dw}(\phi_0)$ and the function

$$f_1(\eta) = -\alpha v \partial_\eta \phi_0(\eta) + \frac{I}{2} g'(\phi_0(\eta)) u_0. \tag{105}$$

Since f_1 is a simple function of η, this equation can formally be solved by inverting the linear operator and applying the result to the function on the right hand side. This operation can only be performed if L_1 is invertible. But it is easy to see that L_1 has an eigenfunction with zero eigenvalue, which is $\partial_\eta \phi_0$. Indeed,

$$\left[\partial_{\eta\eta} - f''_{\rm dw}(\phi_0)\right] \partial_\eta \phi_0 = \partial_\eta \left[\partial_{\eta\eta}\phi_0 - f dw'(\phi_0)\right] = 0, \tag{106}$$

and the quantity between square brackets on the right hand side vanishes because it just corresponds to the equilibrium equation for ϕ_0. This mode corresponds in fact to an infinitesimal displacement of the equilibrium front profile, which cannot give a nontrivial now solution. There is therefore a *solvability condition*: the right hand side of Eq. (104) must be orthogonal to this zero-eigenvalue mode, that is,

$$\int_{-\infty}^{\infty} \partial_\eta \phi_0 f_1(\eta) = 0. \tag{107}$$

This condition fixes the value of u_0. Indeed, after plugging in the expression for f_1 and carrying out the integration, we find

$$\alpha v I + I u_0 = 0, \tag{108}$$

(the chain rule yields $\int \partial_\eta \phi_0 g'(\phi_0) = g(-1) - g(1) = -2$), which simplifies to

$$u_0 = -\alpha v. \tag{109}$$

It can be noticed that, since u_0 is just a constant, this result is exactly equivalent to the calculation of the velocity of the magnetic domain wall given by Eq. (22). The procedure of matched asymptotics thus gives a formal justification for the use of the equilibrium profile in the calculation of this velocity, see Eq. (21).

The next order of the diffusion equation reads

$$v \partial_\eta u_0 = \partial_{\eta\eta} u_1 - \frac{v}{2} \partial_\eta h(\phi_0). \tag{110}$$

Since u_0 is a constant, the left hand side vanishes, and the equation can be integrated once to yield

$$\partial_\eta u_1 - \frac{v}{2} h(\phi_0) + A = 0, \tag{111}$$

where A is a constant of integration. This equation immediately yields the Stefan condition. Indeed, comparing the two limiting behaviors for $\eta \to \pm\infty$ we find

$$\lim_{\eta\to-\infty} \partial_\eta u_1 - \lim_{\eta\to\infty} \partial_\eta u_1 = v \tag{112}$$

Furthermore, since we are interested in a steady-state interface, there is no current far inside the solid (that is, $\partial_\eta u_1 \to 0$ for $\eta \to -\infty$), which fixes $A = v/2$ since $h(\phi_0) \to 1$ for $\eta \to -\infty$. Using this result, Eq. (111) can be integrated once more, which yields

$$u_1 = \bar{u}_1 + \frac{v}{2} \int_0^\eta [h(\phi_0(\xi)) - 1] \, d\xi, \tag{113}$$

where the constant of integration $\bar{u}_1$ can be fixed by the solvability condition for the phase-field equation at the next order,

$$-\alpha v \partial_\eta \phi_1 = \partial_{\eta\eta}\phi_2 + f''_{\mathrm{dw}}(\phi_0)\phi_2 - \frac{I}{2} g'(\phi_0) u_1, \tag{114}$$

which can again be written under the form

$$L_1 \phi_2 = f_2(\eta), \tag{115}$$

where L_1 is the same linear operator as in Eq. (104). In principle, the function ϕ_1 is needed to obtain $f_2(\eta)$. However, simple symmetry arguments show that the contribution of ϕ_1 to the solvability condition vanishes. Indeed, this condition reads

$$\int_{-\infty}^{\infty} \partial_\eta \phi_0 f_2(\eta) = 0. \tag{116}$$

But ϕ_0 is a function that is odd in η ($\phi(-\eta) = -\phi(\eta)$), and thus its derivative is an even function. Therefore, the contribution of any term in f_2 that is odd in η will vanish. Now, ϕ_1 is the solution of Eq. (104) with f_1 given by Eq. (105). Since u_0 is a simple constant, for any function $g(\phi)$ that is odd in ϕ, the right hand side of Eq. (105) is an even function of η. The operator L_1 is also even in η. As a consequence, ϕ_1 must also be even in η, and thus $\partial_\eta \phi_1$ is odd. In fact, several other terms that would be present in a purely formal expansion of the inner phase-field equation have been omitted from Eq. (114) since they do not contribute to the final result due to similar symmetry reasons. In the end, the solvability condition is simply

$$\int \frac{I}{2} g'(\phi_0) u_1(\eta) \partial_\eta \phi_0 \, d\eta = 0, \tag{117}$$

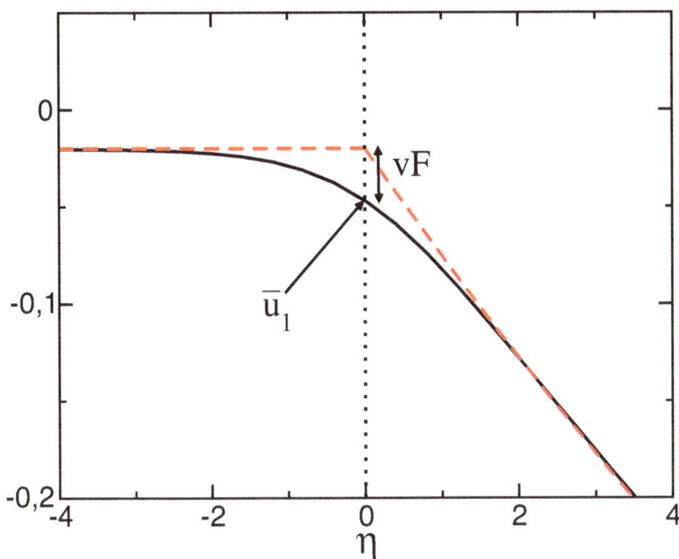

Figure 2. Plot of the temperature field u across a moving planar interface, obtained from a numerical simulation of the model. The dashed lines represent the extrapolations of the asymptotes to the position of the interface, which is indicated by the vertical dotted line. Some of the quantities used in the asymptotic analysis are also indicated: $\bar{u}_1$ is the value of the first-order inner diffusion field at the interface position, and vF is the difference between this value and the "macroscopic" boundary condition.

which yields

$$I\bar{u}_1 = vIJ \tag{118}$$

with J a constant given by

$$J = \int \partial_\eta \phi_0 g'(\phi_0) \left(\int_0^\eta [h(\phi_0(\xi)) - 1] \; d\xi \right) \; d\eta. \tag{119}$$

By this procedure, we have now determined the complete profile of the diffusion field through the interface. However, the boundary condition for the outer field (that is, the value of u that is "seen" on the macroscopic scale) is *not* simply equal to $u_0 + \epsilon\bar{u}_1$. As illustrated in the sketch of Fig. 2, the boundary condition for the outer region is obtained from the application of the matching conditions, Eq. (99), which yields

$$\hat{u}_1|^{\pm} = \bar{u}_1 + vF^{\pm} \tag{120}$$

with

$$F^{\pm} = \int_0^{\pm\infty} [h(\phi_0) - h(\pm 1)] \, d\eta. \tag{121}$$

It is easy to check that for any function $h(\phi)$ odd in ϕ (and in particular for the simplest choice $h(\phi) = \phi$), $F^+ = F^- \equiv F$. This implies that $\hat{u}_1|^+ = \hat{u}_1|^-$, as illustrated in Figure 2. Now we can collect the various pieces of the boundary condition for u, and find

$$u_{\text{int}} = \hat{u}_0 + \epsilon \hat{u}_1 = -v \left(\alpha - \epsilon \frac{J + 2F}{4} \right). \tag{122}$$

Switching back to dimensional variables, and inserting the definition of ϵ, the celebrated expression first obtained by Karma and Rappel (1998) for the interface kinetic coefficient is found,

$$\beta = a_1 \frac{\tau}{\lambda W} \left(1 - a_2 \frac{\lambda W^2}{\tau D} \right), \tag{123}$$

with $a_1 = I/2$ and $a_2 = (J + 2F)/(2I)$. Note that our notations slightly differ from those of Karma and Rappel (1998), and that they use a non-normalized version of the tilting function $g(\phi)$, which leads to small differences between the expressions given here and those of Karma and Rappel (1998).

3.5 Rewiev of some results and discussion

Here, some of the essential consequences of the results obtained above will be discussed, and some examples for their application reviewed; for more details, see Karma and Rappel (1998). As already mentioned before, the first term in Eq. (123) is equivalent to the result for domain wall motion under a homogeneous magnetic field. It can be obtained by a simple calculation in which the temperature is assumed to be *constant* in the entire interface, and has been obtained by several authors in the early days of phase-field modeling, see Langer (1986) or Caginalp (1989). The second term in Eq. (123) arises from the fact that the inhomogeneity of the temperature field across the interface is taken into account. In Karma and Rappel (1998), the asymptotics in which the second term is absent is called *sharp-interface limit*, whereas the full procedure is called the *thin-interface limit*. In the calculation presented here, the thin-interface term appears in a higher order of the asymptotic expansion than the sharp-interface term. This could give the impression as if this term would be subdominant in certain conditions. However, this is incorrect: as explicitly demonstrated by numerical calculations for one-dimensional interfaces in Karma and Rappel (1998), the kinetic behavior of the interface is not correctly described with only the first term in Eq. (123), even in the limit of very slow growth, which corresponds to $\epsilon \to 0$. The physical reason for this behavior is that the heat source term present for moving interfaces has a characteristic scale that is equal to the interface thickness. No matter how small the interface thickness is, there will thus be always a component of the diffusion field that will vary on the scale of the interface. This effect is precisely captured by the above calculation of u_1.

A very important point is that Eq. (123) yields a prescription how to simulate moving interfaces in local equilibrium. This is extremely relevant, since for rough interfaces in metals the interface mobility is large, and thus the kinetic coefficient is small. For the slow growth speeds used in many experiments, kinetic effects can even altogether be neglected, which means that they are adequately described by $\beta = 0$. In the sharp-interface limit, this regime can only be reached when τ tends to zero (which corresponds precisely to a large interface mobility), which leads to equations that are numerically very stiff. The thin-interface expression for β, in contrast, allows for a vanishing kinetic coefficient if W, λ, and τ are related by

$$\tau D = a_2 \lambda W^2. \tag{124}$$

Since $\lambda = a_1 W/d_0$, this is actually a relation between W and τ,

$$\tau = \frac{a_1 a_2}{d_0 D} W^3. \tag{125}$$

Once W is chosen, the above calculation thus fixes the last remaining parameter of the model.

The classic test case for these results is the dendritic growth of a pure substance. To obtain dendrites, anisotropy has to be introduced into the model. Anisotropy of the surface tension can be introduced in the coefficient of the gradient term, or, equivalently, in the interface thickness, whereas an anisotropy in the interface mobility can be introduced in the relaxation time. Both W and τ thus become dependent on the interface orientation,

$$W \to W(\hat{n}) \qquad \tau \to \tau(\hat{n}) \qquad \hat{n} = -\frac{\nabla\phi}{|\nabla\phi|}. \tag{126}$$

Note that the introduction of $W(\hat{n})$ introduces new dependencies on ϕ into the free energy functional and thus generates additional terms in the functional derivative used to obtain the equations of motion, as detailed in Karma and Rappel (1998). These terms implement very naturally the anisotropic version of the Gibbs-Thomson condition, which is quite complicated when expressed in the sharp-interface model. Dendrites are then simulated by starting from a spherical seed in a uniformly undercooled melt. As long as the interface thickness remains at least about an order of magnitude larger than the tip radius, the growth velocity of the tips at steady state, as well as the entire growth transient, are virtually independent of the interface thickness if τ is chosen accoding to the anisotropic generalization of Eq. (125). It should be mentioned that the phase-field model is currently the method of choice for the simulation of dendritic growth. Old predictions have been confirmed and new interesting phenomena have been discovered, see for example Karma et al. (2000) or Haxhimali et al. (2006). The main advantage with respect to sharp-interface models is the simple and natural treatment of the anisotropic Gibbs-Thomson condition.

Another major result is the comparison between the performances of variational and non-variational models. In the results of the matched asymptotics, the two models differ only by the values of a_1 and a_2. Numerical convergence studies presented in Karma and Rappel (1998) demonstrate that the variational model that uses $h(\phi) = g(\phi)$ with $g(\phi)$ given by Eq. (78) requires a finer grid spacing for convergence than the isothermal variational model with $h(\phi) = \phi$. This can be qualitatively understood from the fact that the heat source function is more strongly peaked in the center of the interface for $h(\phi) = g(\phi)$, and therefore a finer grid spacing is necessary to resolve it with sufficient precision. As a result, the computation times of the non-variational model are about one order of magnitude shorter for comparable numerical precision.

This point merits maybe to be restated in a more pronounced way. Going back to the equations that define the phase-field model, we see from Eq. (68) that the local nondimensionalized internal energy density is given by $\tilde{e} = u + g(\phi)/2$. Whereas, in the variatonal model, the evolution equation for u can be transformed into one for $\tilde{e}$ that takes the form of a conservation law, this is no longer possible for the model with $h(\phi) = \phi$. Thererfore, *heat conservation is violated locally* (on the scale of the interface). Loosely speaking, energy is generated at the front side of the interface and destroyed at the rear. This may seem profoundly shocking. However, since the two local violations of energy conservation exactly compensate, on the macroscopic scale the behavior of the model is in perfect agreement with global energy conservation, and it can therefore safely be used as a computational tool for the resolution of the macroscopic free boundary problem. In conclusion, it may sometimes be advantageous from a computational point of view to disrespect thermodynamics, if this is done in a well-controlled manner, as ascertained by the asymptotic analysis.

4 Phase-field models with additions

In this section, I will discuss two examples for phase-field models that obtain a good performance for the solution of given sharp-interface problems by the addition of supplementary terms in the equations of motion that are justified by the results of asymptotic expansions, and that cannot be obtained by variational derivatives of simple free energy functionals. They are examples for the application of phase-field models outside of the framework of the thermodynamic formalism.

4.1 One-sided solidification and antitrapping current

In the previous chapter, the symmetric model of solidification was discussed in detail. This model is usually a quite good approximation for the solidification of a pure substance, in which the relevant diffusion field is the temperature field. However, a situation of far greater practical importance is the solidification of alloys, in which both heat and chemical components must be exchanged between the growing crystal and its environment. A first remark is that the diffusion of heat is usually much faster than chemical diffusion. Therefore, it is often a good approximation to consider *isothermal solidification*, in which the temperature is supposed to be constant and uniform (in practice, it would be set by a heat bath, which can simply be the bulk of a large ingot, for example). Then, crystal growth is controlled by chemical diffusion only. A second important point is that this chemical diffusion is much faster in the liquid than in the solid (by several orders of magnitude). Therefore, the symmetric model is a very bad approximation for this situation. A much better suited model for this case is the *one-sided model*, in which the diffusion inside the solid is altogether neglected.

Whereas, thus, the natural setting for the one-sided model is alloy solidification, the development of equations of motion that include realistic alloy thermodynamics induces some technical difficulties that are not of interest here. Therefore, we will stick with the model for solidification developed above, and just introduce a diffusion coefficient that depends on the phase field and vanishes in the solid. This corresponds to the two-sided thermal model analyzed also by Almgren (1999). It should be kept in mind that, in the framework of alloy solidification, u represents in fact a non-dimensional chemical potential rather than a temperature.

On first glance, it looks as if the modification of the model is very easy. Introducing a diffusion coefficient of the form

$$D(\phi) = D_l q(\phi), \tag{127}$$

where D_l is the diffusivity in the liquid, and $q(\phi)$ is a dimensionless function that satisfies $q(-1) = 1$ (liquid) and $q(1) = 0$ (solid), the equations of motion that seem to be appropriate are simply

$$\tau\partial_t\phi = W^2\nabla^2\phi - f'_{\rm dw}(\phi) - \lambda u g'(\phi), \tag{128}$$

$$\partial_t u = D_l\nabla\left(q(\phi)\nabla u\right) + \frac{1}{2}\partial_t h(\phi), \tag{129}$$

where the only difference to the symmetric model is the presence of $q(\phi)$ in the diffusion equation.

However, the asymptotic analysis quickly shows that matters are not so simple. Nothing changes in the first order of the calculation – the "sharp-interface" terms remain the same as for the symmetric model. But a major difference appears at the next order, in the calculation of the second-order diffusion field u_1. Instead of Eq. (113), we find

$$u_1 = \bar{u}_1 + \frac{v}{2}\int_0^\eta \frac{h(\phi_0(\xi)) - 1}{q(\phi_0(\xi)}\, d\xi. \tag{130}$$

Whereas a plot of this diffusion field looks qualitatively similar to the one obtained from Eq. (113), there is an important difference: the two values of the diffusion field, calculated by application of the matching conditions according to Eq. (120), differ from each other, because the two quantities

$$F^\pm = \int_0^{\pm\infty}\left[\frac{h(\phi_0) - 1}{q(\phi_0)} - \frac{h(\pm 1) - 1}{q(\pm 1)}\right] d\eta. \tag{131}$$

are *not* equal in the one-sided model. As a result, the macroscopic diffusion field exhibits a jump at a moving interface that is proportional to the interface velocity v, the difference $F^+ - F^-$, and to the interface thickness W. The latter point can be appreciated after switching back from dimensionless to dimensional variables in the calculation of u_1. An example of such a jump is shown in Fig. 3. It should be stressed that there is *no discontinuity* in the field u on the scale of the interface – this discuntinuity only appears when the system in seen on the macroscopic scale.

In the framework of alloy solidification, this jump corresponds to the phenomenon of *solute trapping*, which generates indeed a discontinuity of the chemical potential at the interface. It arises from the fact that interface propagation can be much faster than the characteristic speed for exchange of chemical components through the interface. In the classic continuum theory of Aziz (1982), this effect is indeed proportional to both V and W. Therefore, the phase-field model makes a prediction that is qualitatively correct. However, trouble arises if this model is to be used as a computational tool

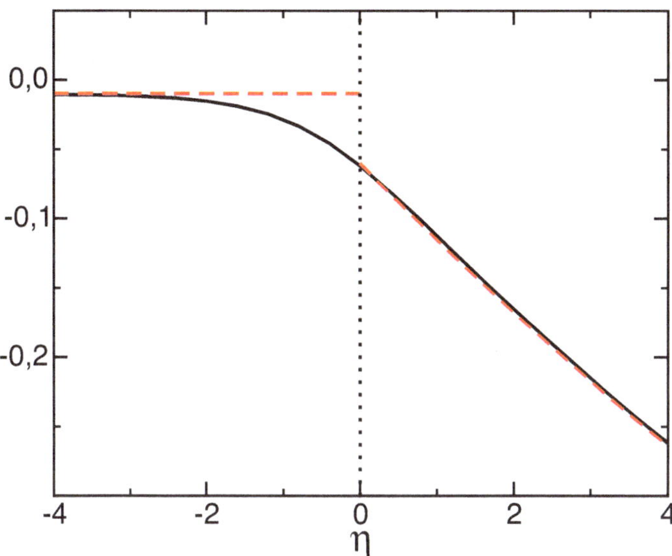

Figure 3. Plot of the temperature field u across a moving planar interface, obtained from a numerical simulation of the one-sided model model. The two extrapolated asymptotes (dashed lines) do not yield the same value of u at the interface (vertical dotted line) on the liquid and the solid side.

in which the interface thickness W is upscaled, because then the magnitude of this effect will be greatly increased, which makes it appreciable even at growth speed where it is completely negligible for the "physical" value of the interface thickness.

In a search how to cure this problem, the one-dimensional calculation performed here cannot give a complete answer. Indeed, an obvious idea is to modify the interpolation functions used in the model to restore the equality $F^+ = F^-$, which makes the solute trapping effect disappear. One possibility to do this that can be deduced from Eq. (131) is to choose $h(\phi)$ in such a way that $(h(\phi)-1)/q(\phi) = \phi - 1$, which would make the expressions of $F^\pm$ identical to those of the symmetric model with $h(\phi) = \phi$. However, the full two-dimensional calculations carried out in Almgren (1999) and Echebarria et al. (2004) reveal that there are two additional constraints. The reason why these constraints appear only in two dimension is that they imply curvature and diffusion along the interface, respectively. The first concerns the heat source function $h(\phi)$. Consider a curved moving interface. Its length will change with time at a local rate that is given by the product

$v\kappa$. Furthermore, for a curved interface the lengths of the "forward" and "backward" side of the interface are not equal. Therefore, the integration of the heat source term through a moving curved interface gives rise to an excess heat production that is proportional to the interface stretching rate $v\kappa$ and to $H^+ - H^-$, where

$$H^{\pm} = \int_0^{\pm\infty} [h(\phi_0) - h(\pm 1)] \; d\eta. \tag{132}$$

The second correction term concerns diffusion *along* the interface. Indeed, consider a gradient of the diffusion field u along an interface. Since the diffusion coefficient varies through the interface, diffusion will be faster than in the solid on the solid side of the interface, and slower than in the liquid on the liquid side. This will give rise to an additional *surface diffusion* term in the macroscopic equations, unless the two excesses exatcly compensate, that is $Q^+ = Q^-$ with

$$Q^{\pm} = \int_0^{\pm\infty} [q(\phi_0) - q(\pm 1)] \; d\eta. \tag{133}$$

Like the solute trapping effect, both of these additional terms are proportional to the interface thickness. In a quantitatively accurate model, all three need therefore to be eliminated. As already shown in Almgren (1999), it is not possible to achieve this with a reasonable choice of interpolation functions. The new idea put forward by Karma (2001) and further elaborated in Echebarria et al. (2004) is to add a new term, the so-called *antitrapping current* to the diffusion equation in order to counterbalance solute trapping. Since solute trapping acrually corresponds to an engulfment of solute atoms in the advancing interface, this antitrapping current should be directed in the direction normal to the interface and from the solid to the liquid, and be proportional to the interface velocity. Therefore, a reasonable choice is

$$j_{\mathrm{at}} = a(\phi) W \hat{n} \dot{\phi}, \tag{134}$$

where $a(\phi)$ is a new interpolation function, and $\dot{\phi}$ denotes the time derivative of the phase field. With this addition, the diffusion equation for u becomes

$$\partial_t u = D_l \nabla \left(q(\phi) \nabla u \right) - \nabla \cdot j_{\mathrm{at}}. \tag{135}$$

This new ingredient indeed makes it possible to restore the macroscopic continuity of the diffusion field through the interface. Indeed, the calculation of u_1 now yields

$$u_1 = \bar{u}_1 + \frac{v}{2} \int_0^{\eta} p(\phi_0(\xi)) \; d\xi \tag{136}$$

with the function $p(\phi_0)$ given by

$$p(\phi_0) = \frac{h(\phi_0) - 1 - 2a(\phi_0)\partial_\eta \phi_0}{q(\phi_0)}. \tag{137}$$

For the simplest model with $h(\phi) = \phi$ and $q(\phi) = (1-\phi)/2$, the choice of a constant $a(\phi) = 1/(2\sqrt{2})$ yields $p(\phi) = \phi - 1$ and therefore restores a profile of u_1 that is identical to the one of the symmetric model. With this trick, the rest of the asymptotic calculation is unchanged, and the same expression for the kinetic coefficient is obtained than in the symmetric model.

It should be stressed that up to now, no possibility has been found to obtain the antitrapping current from a variational formulation. But it performs very well in numerical calculations, as demonstrated in benchmark simulations in Echebarria et al. (2004). Moreover, up to now no method has been found to formulate a phase-field model that is free from thin-interface effects for an arbitrary ratio of the two diffusivities. As long as the diffusivity in the solid remains much smaller than the one in the liquid, the approach developed for the one-sided model continues to work well. For arbitrary diffusivity ratios, a generalization of the antitrapping current has been developed by Ohno and Matsuura (2009). It works well in situations where the diffusion current on the solid side of the interface remains small. However, as shown by Plapp (2011), in the presence of a non-vanishing current that crosses the interface, a new thin-interface effect appears (interface resistance). So far, it is unclear how this additional effect can be eliminated. In the absence of a solution to this problem, there is no general model for two-phase solidification that has the same efficiency as the ones for the symmetric and the one-sided model.

4.2 The advected field method

The final example for a phase-field model to be discussed is a model for two-phase fluid flow of two immiscible fluids. This is a classic subject in fluid mechanics, and numerous diffuse-interface and phase-field models for this problem have been developed – for a review, see Anderson et al. (1998). Classic approaches are the second gradient method, discussed for example by Jamet et al. (2001), and the convective Cahn-Hilliard equation, see Jacqmin (1999) and elsewhere in the present volume *Roberto: put cross-reference to the part by Anderson.*

An approach that uses the insights from asymptiotic analysis in the spirit of the preceding sections was developed by Folch et al. (1999) for two-phase flow in a Hele-Shaw cell, and later generalized to arbitrary two-phase flows by Biben et al. (2003), who coined the name *advected field method.* The general idea is that for immiscible fluids, there is no mass exchange across the interface, which is therefore advected by the flow. Furthermore, if the fluids are incompressible, the normal component of the fluid velocity is continous across the interface. Then, the geometry of the fluid domains should be described by a phase-field function that acts as a "marker field", which has no dynamics on its own. This can be achieved with the evolution equation

$$\partial_t \phi + v \cdot \nabla \phi = \Gamma \left[W^2 \left(\nabla^2 \phi - \kappa(\phi) |\nabla \phi| \right) - f'_{\mathrm{dw}}(\phi) \right]. \tag{138}$$

Here, v is an arbitrary velocity field that is the solution of the appropriate fluid flow equation (see below), Γ is a constant, W the interface thickness, and $\kappa(\phi)$ is the local curvature given by

$$\kappa(\phi) = \nabla \cdot \hat{n} = \nabla \cdot \frac{\nabla \phi}{|\nabla \phi|}. \tag{139}$$

The left hand side of Eq. (138) is just the advected derivative of the scalar ϕ. The right hand side without the second term is identical to the expression obtained for the time evolution of m in the magnetic model (model A), see Eq. (20). To understand the role of the additional term, it is useful to analyze what happens to a spherical inclusion of the phase described by $\phi = -1$ inside an infinite matrix of the $\phi = 1$ phase. In this situation, $\nabla \phi = \partial_r \phi$, and $\kappa(\phi)$ is positive according to Eq. (139) since $\partial_r \phi > 0$. Furthermore, the Laplacian written in spherical coordinates is simply $\nabla^2 = \partial_{rr} + ((d-1)/r)\partial_r$. Therefore, Eq. (138) becomes

$$0 = \left[W^2 \left(\partial_{rr} \phi + \frac{d-1}{r} \partial_r \phi - \kappa(\phi) |\nabla \phi| \right) - f'_{\mathrm{dw}}(\phi) \right]. \tag{140}$$

But since $\kappa(\phi) = (d-1)/r$ for any spherically symmetric inclusion, the additional term exactly cancels the $1/r$ term of the Laplacian, such that this equation becomes exactly the equilibrium equation of a planar interface. Thus, the new term $\kappa(\phi)|\nabla \phi|$ is actually a *counterterm* which exactly cancels the driving force for motion by curvature. As a result, the spherical inclusion that would disappear in finite time in the magnetic model is actually an equilibrium configuration here.

This equation was originally developed based on calculations similar to those presented in the previous chapter. Several years later, it was shown by Jamet and Misbah (2008) that, quite surprisingly, this equation can be obtained as the variational derivative of a free energy functional. However, this functional is more complicated than the functionals displayed so far: in addition to the gradient and potential terms present in all the functionals discussed up to now, it contains an additional term that mixes gradients and the local values of the fields. If this term is chosen in the right way, it guarantees the existence of front solutions *without excess free energy* by creating orbits in the "phase space" of the model (spanned by the values of ϕ and $\nabla\phi$) along which the free energy density is constant and exactly equal to its value at the potential minima. This yields a natural alternative explanation for the properties of this equation: without excess free energy (surface tension), there is indeed no driving force for motion by curvature.

This equation for the phase field has now to be coupled with an equation for fluid flow. As an example, let us take the Navier-Stokes equations for Newtonian fluids,

$$\rho(\phi)\left(\partial_t v + v \cdot \nabla v\right) = \nabla \cdot \sigma_{ij} - \nabla P - \kappa(\phi)\sigma\frac{\nabla\phi}{2}. \qquad (141)$$

Here, $\rho(\phi)$ is the phase-dependent density, P is the pressure, σ (without indices) is, as before, the surface tension, whereas $\sigma_{ij} = \eta(\phi)(\partial_i v_j + \partial_j v_i)$ is the Newtonian dissipative stress tensor, with $\eta(\phi)$ the phase-dependent viscosity. The only term which merits further explanations is the last one, which actually represents the capillary force created by the surface tension. Indeed, as usual in phase-field models, this surface force has been transformed into a volumetric force that is "smeared out" over the diffuse interface. For constant curvature κ, an integration of this term through the interface just yields $\kappa\sigma$ since ϕ varies from -1 to 1. Therefore, we recover the Laplace law for the pressure difference between the two sides of a curved interface. The density and the viscosity are interpolated in the usual way, for example for the viscosity we have $\eta(\phi) = \eta_1(1+\phi)/2 + \eta_2(1-\phi)/2$, where η_1 and η_2 are the viscosities of the two fluids.

There are at least two advantages of this approach with respect to the convective Cahn-Hillard equation. First, as discussed in Sec. 2.2, the Cahn-Hilliard equation exhibits Ostwald ripening: when a large and a small droplet are present simultaneously, the small one evaporates and its materials joins the larger one by diffusion. Whereas this is a realistic process, for immiscible fluids the mutual solubilities of the two substances are so low that this process would be extremely slow. The advected field method handles this situation without difficulties: two well-separated drops will remain stationary, whatever are their respective sizes. Second, since the advected field method can be coupled with any fluid flow equation, it is quite easy to extend it to non-Newtonian fluids or even more complex situations. For instance, it has been used for shear-thinning fluids by Nguyen et al. (2010) and for viscoelastic fluids by Beaucourt et al. (2005), as well as (with some modifications) for the time evolution of vesicles by Biben et al. (2005).

5 Summary and Conclusions

In this course, I have presented how phase-field models can be obtained, on the one hand, from the description of the dynamics of first-order phase transitions, and one the other hand from the regularization of macroscopic free-boundary problems. The most important messages can be summarized as follows:

- The models in which interfaces are described as the profile of a physically accessible quantity are easy to obtain from Landau expansions and mean-field arguments, but have numerous computational constraints due to the fact that both the bulk and interface properties of the order parameters are determined by the same free energy curve.
- Phase-field models can alternatively be obtained by starting from the bulk thermodynamics and supplementing it with a dynamical phase field that plays the role of a smoothed indicator function. For a judicious choice of the functions that interpolate the physical quantities between the phases, interface and bulk properties can be chosen independently from each other.
- The link between phase-field models and free-boundary problems can be established quite rigorously by the method of matched asymptotic expansions. As shown for the case of solidification, a correct description of the interface dynamics can only be obtained when *all* phenomena taking place on the scale of the interface thickness are taken into account.
- While it is formally appealing to obtain the equations of motion from a variational formalism, this is not necessary if the goal is to obtain a phase-field formulation for a prescribed free boundary problem. Sometimes, non-variational models may even have large computational advantages.
- Once the asymptotic behavior of a model is well understood, certain properties can be adjusted by adding new terms to the equations of motion.

Only a few simple and fundamental problems that can be solved with phase-field models have been discussed here. The improvement of existing models, and the development of new phase-field models for different physical phenomena is a widely open and very active field of research. An overview over many other applications is given, for example, in the review articles by Emmerich (2008); Steinbach (2009); Wang and Li (2010), and also in several other contributions to the present volume.

Acknowledgments: I thank Tuan Pham Ngoc for a careful proofreading of this manuscript.

Bibliography

R. F. Almgren. Second-order phase field asymptotics for unequal conductivities. *SIAM J. Appl. Math.*, 59:2086–2107, 1999.

D. M. Anderson, G. B. McFadden, and A. A. Wheeler. Diffuse-interface methods in fluid mechanics. *Annual Review of Fluid Mechanics*, 30:139, 1998.

D. M. Anderson, G. B. McFadden, and A. A. Wheeler. A phase-field model of solidification with convection. *PHYSICA D*, 135:175–194, 2000.

M. J. Aziz. Model for the solute redistribution during rapid solidification. *J. Appl. Phys.*, 53(2):1158, 1982.

J. Beaucourt, T. Biben, and C. Verdier. Elongation and burst of axisymmetric viscoelastic droplets: A numerical study. *Phys. Rev. E*, 71:066309, 2005.

C. Beckermann, H.-J. Diepers, I. Steinbach, A. Karma, and X. Tong. Modeling melt convection in phase-field simulations of solidification. *J. Comput. Phys.*, 154:468, 1999.

T. Biben, C. Misbah, A. Leyrat, and C. Verdier. An advected-field approach to the dynamics of fluid interfaces. *Europhys. Lett.*, 63:623, 2003.

T. Biben, K. Kassner, and C. Misbah. Phase-field approach to three-dimensional vesicle dynamics. *Phys. Rev. E*, 72:041921, 2005.

W. J. Boettinger, J. A. Warren, C. Beckermann, and A. Karma. Phase-field simulation of solidification. *Annu. Rev. Mater. Res.*, 32:163–194, 2002.

A. J. Bray. Theory of Phase-ordering kinetics. *Adv. Phys.*, 43:357–459, 1994.

Q. Bronchart, Y. Le Bouar, and A. Finel. New coarse-grained derivation of a phase field model for precipitation. *Phys. Rev. Lett.*, 100:015702, 2008.

G. Caginalp. Stefan and hele-shaw type models as asymptotic limits of the phase-field equations. *Phys. Rev. A*, 39:5887–5896, 1989.

J. W. Cahn and J. E. Hilliard. Free energy of a non-uniform system. 1. interfacial free energy. *J. Chem. Phys.*, 28:258–267, 1958.

L.-Q. Chen. Phase-field models for microstructure evolution. *Annu. Rev. Mater. Res.*, 32:113, 2002.

J. B. Collins and H. Levine. Diffuse interface model of diffusion-limited crystal growth. *Phys. Rev. B*, 31:6119–6122, 1985.

B. Echebarria, R. Folch, A. Karma, and M. Plapp. Quantitative phase-field model of alloy solidification. *Phys. Rev. E*, 70(6):061604, 2004.

K. R. Elder, M. Grant, N. Provatas, and J. M. Kosterlitz. Sharp interface limits of phase-field models. *Phys. Rev. E*, 64:021604, 2001.

H. Emmerich. Advances of and by phase-field modelling in condensed-matter physics. *Adv. Phys.*, 57(1):1–87, 2008.

G. J. Fix. In A. Fasano and M. Primicerio, editors, *Free boundary problems: Theory and applications*, page 580, Boston, 1983. Piman.

R. Folch, J. Casademunt, A. Hernândez-Machado, and L. Ramírez Piscina. Phase-field model for hele-shaw flows with arbitrary viscosity contrast. i. theoretical approach. *Phys. Rev. E*, 60(2):1724, 1999.

M. E. Glicksman, M. B. Koss, and E. A. Winsa. Dendritic growth velocities in microgravity. *Phys. Rev. Lett.*, 73:573–576, 1994.

T. Haxhimali, A. Karma, F. Gonzales, and M. Rappaz. Orientation selection in dendritic evolution. *Nature Materials*, 5:660–664, 2006.

P. C. Hohenberg and B. I. Halperin. Theory of dynamic critical phenomena. *Rev. Mod. Phys.*, 49:435–479, 1977.

J. J. Hoyt, M. Asta, and A. Karma. Atomistic and continuum modeling of dendritic solidification. *Mat. Sience Eng. R*, 41:121, 2003.

D. Jacqmin. Calculation of two-phase Navier-Stokes flows using phase-field modeling. *J. Comput. Phys.*, 155:96–127, 1999.

D. Jamet and C. Misbah. Thermodynamically consistent picture of the phase-field model of vesicles: Elimination of the surface tension. *Phys. Rev. E*, 78:041903, 2008.

D. Jamet, O. Lebaigue, N. Coutris, and J. M. Delhaye. The second gradient method for the direct numerical simulation of liquid-vapor flows with phase change. *J. Comput. Phys.*, 169:624–651, 2001.

A. Karma. Phase-field formulation for quantitative modeling of alloy solidification. *Phys. Rev. Lett.*, 87(10):115701, 2001.

A. Karma and W.-J. Rappel. Phase-field method for computationally efficient modeling of solidification with arbitrary interface kinetics. *Phys. Rev. E*, 53(4):R3017–R3020, 1996.

A. Karma and W.-J. Rappel. Quantitative phase-field modeling of dendritic growth in two and three dimensions. *Phys. Rev. E*, 57(4):4323–4349, 1998.

A. Karma, Y. H. Lee, and M. Plapp. Three-dimensional dendrite-tip morphology at low undercooling. *Phys. Rev. E*, 61(4, Part B):3996–4006, APR 2000.

J. S. Langer. An introduction to the kinetics of first-order phase transitions. In C. Godrèche, editor, *Solids far from equilibrium*, Edition Aléa Saclay, pages 297–363, Cambridge, UK, 1991. Cambridge University Press.

J. S. Langer. Models of pattern formation in first-order phase transitions. In G. Grinstein and G. Mazenko, editors, *Directions in Condensed Matter Physics*, pages 165–186, Singapore, 1986. World Scientific.

S. Nguyen, R. Folch, V. K. Verma, H. Henry, and M. Plapp. Phase-field simulations of viscous fingering in shear-thinning fluids. *Phys. Fluids*, 22:103102, 2010.

M. Ohno and K. Matsuura. Quantitative phase-field modeling for dilute alloy solidification involving diffusion in the solid. *Phys. Rev. E*, 79(3): 031603, 2009.

O. Penrose and P. C. Fife. Thermodynamically consistent models of phase-field type for the kinetics of phase transitions. *Physica D*, 43:44–62, 1990.

M. Plapp. Three-dimensional phase-field simulations of directional solidification. *J. Cryst. Growth*, 303:49–57, 2007.

M. Plapp. Remarks on some open problems in phase-field modelling of solidification. *Phil. Mag.*, 91:25–44, 2011.

J. S. Rowlinson. Translation of J. D. van der Waals, The Thermodynamic Theory of Capillarity under the Hypothesis of a Continuous Variation of Density. *J. Stat. Phys.*, 20:197–244, 1979.

I. Steinbach. Phase-field models in materials science. *Model. Simul. Mater. Sci. Eng.*, 17(7):073001, OCT 2009.

W. van Saarloos. Front propagation into unstable states. *Phys. Reports*, 386:29–222, 2003.

Y. Wang and J. Li. Phase field modeling of defects and deformation. *Acta Mater.*, 58:1212–1235, 2010.

Zeitfracht Medien GmbH
Ferdinand-Jühlke-Straße 7
99095 Erfurt, Deutschland
produktsicherheit@kolibri360.de